1,000,000 Books

are available to read at

Forgotten Books

www.ForgottenBooks.com

Read online
Download PDF
Purchase in print

ISBN 978-1-332-31783-7
PIBN 10313228

This book is a reproduction of an important historical work. Forgotten Books uses
state-of-the-art technology to digitally reconstruct the work, preserving the original format
whilst repairing imperfections present in the aged copy. In rare cases, an imperfection in
the original, such as a blemish or missing page, may be replicated in our edition. We do,
however, repair the vast majority of imperfections successfully; any imperfections that
remain are intentionally left to preserve the state of such historical works.

Forgotten Books is a registered trademark of FB &c Ltd.
Copyright © 2018 FB &c Ltd.
FB &c Ltd, Dalton House, 60 Windsor Avenue, London, SW19 2RR.
Company number 08720141. Registered in England and Wales.

For support please visit www.forgottenbooks.com

1 MONTH OF FREE READING

at
www.ForgottenBooks.com

By purchasing this book you are eligible for one month membership to ForgottenBooks.com, giving you unlimited access to our entire collection of over 1,000,000 titles via our web site and mobile apps.

To claim your free month visit:
www.forgottenbooks.com/free313228

* Offer is valid for 45 days from date of purchase. Terms and conditions apply.

English
Français
Deutsche
Italiano
Español
Português

www.forgottenbooks.com

Mythology Photography **Fiction** Fishing Christianity **Art** Cooking Essays Buddhism Freemasonry Medicine **Biology** Music **Ancient Egypt** Evolution Carpentry Physics Dance Geology **Mathematics** Fitness Shakespeare **Folklore** Yoga Marketing **Confidence** Immortality Biographies Poetry **Psychology** Witchcraft Electronics Chemistry History **Law** Accounting **Philosophy** Anthropology Alchemy Drama Quantum Mechanics Atheism Sexual Health **Ancient History Entrepreneurship** Languages Sport Paleontology Needlework Islam **Metaphysics** Investment Archaeology Parenting Statistics Criminology **Motivational**

PRACTICAL TALKS ON FARM ENGINEERING

*A Simple Explanation of Many Everyday
Problems in Farm Engineering and Farm
Mechanics Written in a Readable
Style for the Practical Farmer*

BY

R. P. CLARKSON, B. S.

PROFESSOR OF ENGINEERING, ACADIA UNIVERSITY
CONSULTING ENGINEER
ENGINEERING CORRESPONDENT OF THE "RURAL NEW-YORKER"

Illustrated from photographs and diagrams

GARDEN CITY   NEW YORK
DOUBLEDAY, PAGE & COMPANY

Copyright, 1915, by
DOUBLEDAY, PAGE & COMPANY

*All rights reserved, including that of
translation into foreign languages,
including the Scandinavian*

ACKNOWLEDGMENTS

The author and publishers wish to extend to the following firms acknowledgment and thanks for their hearty coöperation:

AVERY COMPANY, PEORIA, ILLINOIS

CANADA CEMENT COMPANY, MONTREAL, QUEBEC

CROSBY STEAM GAUGE AND VALVE COMPANY, BOSTON, MASS.

DETROIT ENGINE COMPANY, DETROIT, MICH.

DODD AND STRUTHERS, DES MOINES, IOWA

HOLT TRACTOR COMPANY, STOCKTON, CAL.

DUNT MOSS COMPANY, BOSTON, MASS.

AUTHOR'S PREFACE

THE information set forth in this little volume is gleaned from the experience of a number of years spent in advising and aiding farmers in these matters, and is largely an outgrowth of the material I have printed in the *Rural New-Yorker* in reply to hundreds of questions from its readers in all portions of America. In a few cases the "talk" is based on articles contributed by me to prominent farm journals of both the United States and Canada, such as *Hoard's Dairyman, The Field Illustrated, American Cultivator, Northwestern Agriculturist, Kimball's Dairy Farmer, Up-to-date Farming,* of the United States, and *The Farmer's Advocate, The Weekly Sun, The Mail and Empire,* and the *Manitoba Free Press,* of Canada. In every case, however, the material has been carefully revised and rewritten and new illustrations used. There has been no attempt whatever to make this a textbook or even a treatise on Farm Engineering. Its sole aim is to present the material in an interesting and popular form for the use of farmers who have neither the training nor the inclination to wade through more technical volumes on the subject. The list of topics included is not a long one, nor does it by any means exhaust the interests of the farmers. It is a selection made from the things most often inquired about during the last three or four years and found to be most perplexing to the farmers.

R. P. CLARKSON.

THE FIELD OF FARM ENGINEERING

As ONE thinks over the work of the farmer, it is astonishing to note how much engineering enters into it. The choice of materials for buildings, for roads, walks, and fences; the erection of the buildings themselves; the care and operation of machinery, including tractors and automobiles; the selection of a good water supply and the system whereby the water is made available for domestic purposes, for stock, and for crop irrigation; the drainage of land; the disposal of sewage; the installation of farm power, possibly by harnessing some small, swiftly running stream or some waterfall; the industrial use of crops and the waste products from crops—all this is engineering work and lies rather in the province of the engineer than in that of the agriculturist alone.

These are merely the general engineering subjects with which every farmer deals. Many cases arise where special engineering information is of value. Trouble with the telephone lines; the matter of lightning protection; the value and disposal of water and mineral rights; the utilization of raw products found in the land, such as limestone, coal, peat, oil, metal ores, and similar substances; even the operation of the furnace and the choice of fuel are subjects which receive careful study in any first-class engineering training.

Farming might almost be defined as a branch of engineering if we take the generally accepted definition that engineering is the development of the resources of nature for the use and convenience of man.

TABLE OF CONTENTS

PART I

FARM BUILDINGS AND BUILDING MATERIALS

	PAGE
Farm Building, Design, and Construction	3
The Farm Icehouse	11
The Principles of Cold Storage	21
The Waterproofing of Concrete	26
Artificial Stones and Composition Flooring	36
Paints and Painting	43
Lightning Rods and Rodding	47

PART II

FARM WATER SUPPLY AND SEWAGE DISPOSAL

The Sources of a Pure Water Supply	61
Running Water for Fifteen Dollars	68
A Sand Filter for Rain or Brook Water	75
Softening Hard Water	79
The Hydraulic Ram and the Ram-pump	82
Disposal of House Sewage	89

PART III

FARM POWER

Kerosene, Gasoline, and Coal as Fuels	97
The Oil Tractor on the Small Farm	102

	PAGE
The Ignition System and Ignition Control of the Gasoline Engine	111
Determining the Horsepower of an Engine	119
Utilizing Small Streams for Power	129
The Storage Battery for the Farm	145

PART IV

DRAINAGE AND IRRIGATION

The Principles of Drainage	159
Construction of the Tile Drain	172
Some Facts Concerning Small Irrigation Practice	178

PART V

MISCELLANEOUS ENGINEERING TALKS

The Cost of Road Building	189
The Working Principles of Orchard Heaters	194
The Forms of Electricity	199

PART VI

USEFUL TABLES FOR ENGINEERING CALCULATIONS

I.	The Equivalents of One Horsepower	205
II.	Absolute Efficiency of Various Engines	206
III.	Weights of Various Materials	208
IV.	Strength of Various Materials	212
V.	The Heating Value of Fuels	214
VI.	Water Heads and Corresponding Pressures	216
VII.	Water Powers for Various Heads	217

INDEX 219

LIST OF ILLUSTRATIONS

Line Cuts and Half-tones

FIGURE		FACING PAGE
20.	THE CATERPILLAR TRACTOR ADAPTED FOR SOFT SOILS PARTICULARLY	*Frontispiece*
1.	RELATIVE LENGTHS OF THE ENCLOSED LINES FOR THE THREE EQUAL AREAS A, B, AND C—*In Text*	5
2.	APPROACHING THE ROUND CONSTRUCTION. NOTE DRIVEWAY AT LEFT TO THE MIDDLE FLOOR	6
3.	TAKING ADVANTAGE OF THE ECONOMY WHICH RESULTS FROM THE ROUND CONSTRUCTION	7
4.	DIAGRAM SHOWING VENTILATING SYSTEM—*In Text*	7
5.	THE TALL, ROUND SILO IS BEST	8
6.	AN EXAMPLE OF TRUE BUILDING EFFICIENCY	9
7.	SIMPLE ROOF VENTILATORS FOR ICEHOUSE CONSTRUCTION—*In Text*	18
8.	SHOWING THE DISCHARGE BETWEEN CLOUDS AND THE OVERFLOW TO EARTH	48
9.	LIGHTNING PROTECTION FOR THE ROOF VENTILATOR—*In Text*	50
10.	THE METHOD OF BENDING BARBED WIRE TO FORM AN ENCLOSING NETWORK FOR LIGHTNING PROTECTION—*In Text*	55

FIGURE		FACING PAGE
11.	THE PROPER ARRANGEMENT FOR THE TOP OF A DUG WELL—*In Text*	63
12.	A NEAT AND DESIRABLE SPRING HOUSING	70
13.	A SIMPLE RUNNING WATER SYSTEM OF LOW COST—*In Text*	69
14.	A SIMPLE PNEUMATIC EQUIPMENT—*In Text*	70
15.	ILLUSTRATING THE SIMPLICITY OF THE HOT-WATER SYSTEM—*In Text*	72
16.	A SATISFACTORY SAND FILTER—*In Text*	77
17.	DIAGRAM SHOWING PARTS OF RAM AND RAM-PUMP—*In Text*	85
18.	THE SEPTIC TANK—*In Text*	90
19.	A TRACTOR IN THE LUMBER COUNTRY	102
21.	A CATERPILLAR TRACTOR WORKING IN GROUND AFTER PLOWING	103
22.	A CATERPILLAR TRACTOR WORKING IN SWAMP LAND	106
23.	THE PAST AND THE PRESENT	107
24.	A SEVERE TEST FOR ANY MACHINE	107
25.	THE ONE MAN OUTFIT PLOWING	108
26.	THE ENGINE INDICATOR—*In Text*	121
27.	A TYPICAL INDICATOR CARD—*In Text*	123
28.	ONE FORM OF PRONY BRAKE—*In Text*	124
29.	TYPES OF WATER-WHEELS—*In Text*	133
30.	DIAGRAMMATIC REPRESENTATION OF TYPICAL TURBINE WHEELS—*In Text*	138
31.	GENERAL LOCATION OF DAM AND TURBINE WHEEL IN MOST INSTALLATIONS—*In Text*	140
32.	CHARGING A SMALL STORAGE BATTERY WITH ALTERNATING CURRENT LIGHTING CIRCUIT	146
33.	THE GRID BEFORE PASTING	147
34.	THE PASTED PLATE COMPLETED	147

FIGURE		FACING PAGE
35.	The Planté Plate After Shredding but Before "Forming" the Paste	147
36.	The Planté Plate Completed and "Formed"	147
37.	The Lead Cell at the Left. The Edison at the Right	150
38.	The Edison Positive Plate	151
39.	The Edison Negative Plate	151
40.	A Kerosene Engine Belted to a Lighting Generator	154
41.	A Natural System of Drainage—*In Text*	167
42.	Other Drainage Systems—*In Text*	170
43.	An Orchard Heater—*In Text*	196

PART I

FARM BUILDINGS AND BUILDING MATERIALS

Farm Building, Design, and Construction.
The Farm Icehouse.
The Principles of Cold Storage.
The Waterproofing of Concrete.
Artificial Stones and Composition Flooring.
Paints and Painting.
Lightning Rods and Rodding.

PRACTICAL TALKS ON FARM ENGINEERING

CHAPTER I

Farm Building, Design, and Construction

There are a great many important small things which everybody knows when he has leisure to recall them, but which do not always come to mind at just the time when they are most needed. In planning for the future care and repair of the new farm buildings it is often the little things that count. We must build not only with an eye to present conditions but also keeping in mind the frequent renewal of some part of the building, the painting and upkeep and, more important still, the ease and convenience with which we shall be able to work in the building. There is no greater mistake made than to consider a thing which makes work easier as a luxury. It is a prime necessity, because anything which makes work easier will make the worker able to do more and thus will make what he does do cost less. If money is at hand to make this possible, every possible time saver should be employed. Consider, for example, a plan which would save two hours a week in the care of horses or stock. That means one hundred and four hours a year could be used in doing some other work, and at fifteen cents an hour, the sum of $15.60 could be earned. This amount is the interest on three hundred and twelve dollars at 5 per cent. That is, the plan which saves two hours a week of your time gives you as much cash return in a year as would the sum of $312 invested in the bank. Therefore that time-saving plan is worth $312 to you.

In planning the shape of buildings, keep in mind that you wish to enclose the largest possible space for the least amount of money. A round building will do this but a square building is a close second choice. The most wasteful shapes are the long narrow buildings. For example, consider a building built round and 77 feet in diameter. The space enclosed will be about 4,660 square feet and the length of the wall along the ground will be 242 feet. A square building enclosing the same space will have to be slightly more than 68 feet square. Thus the wall will be 272 feet long, so that, if the barns are the same height, the square barn will take about 12 per cent. more lumber for building the walls, and every time it is painted it will cost 12 per cent. more to do it. The round barn will be warmer in winter because

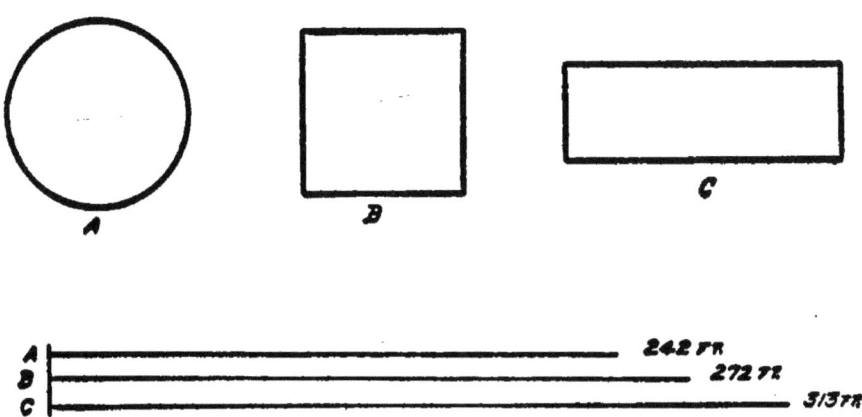

Fig. 1.—Relative lengths of the enclosing lines for the three equal areas A, B, and C

there is less radiating wall space. For the same reason it will be cooler in summer. The saving of either of these barns over a long barn is very great. To enclose the same space as these, suppose we build a barn 116½ feet long by 40 feet wide. The wall space would be 313 feet long or nearly 30 per cent. more lumber and paint would be required than in the case of the round barn, and about 15 per cent. more than if the square barn were built. Such a long barn would be colder in winter and warmer in summer.

The round barn of the "consolidated" type with a round silo in the centre and all the animals and feed under one roof is nearest to the ideal construction. Less time and labour are required in caring for the stock, with the result that better care is given them. Make it a gravity barn, as I like to call it. That is, have a driveway to the upper floor so that hay, grain, and feed may be delivered up there, and manure, waste, etc., taken out at the bottom floor. Have all the movement of material done by gravity. If you want a thrashing-floor, have it on the top floor where the wagons can unload directly. Then have the granary on the floor below so that the grain will not have to be lifted. Put the stock on the floor below this and have the feed go down to them through a chute. If grain is to be shipped, the wagons may be loaded at the lowest floor. In fact, anything in the barn may be loaded in wagons without lifting.

A three floor barn is by no means impracticable, nor is it an impossibility in any location

Fig. 2.—Approaching the round construction. Note driveway at left to the middle floor. This barn is entirely of concrete. It is 12 sided, 60 feet in diameter, each side 16 feet long. Main walls 30 feet high, 12 inches thick for 10 ft. up, 10 inches for the next 10 ft, and 8 inches for the rest. Cost about $2,000, complete

FIG. 3.—Taking advantage of the economy which results from the round construction. This barn is 100 feet long, 50 feet wide, and 40 feet from sill to peak. Note ventilators over windows and below sheathing. Cost of building about $7,000

whatever. Such barns are frequent, but of course the construction of proper driveways is much easier where the barn can be built against or near a hillside or bank. If some of these arrangements must be changed to fit your case, use what you can, for the nearer you come to the ideal construction, the more convenient and therefore the more valuable your barn will be.

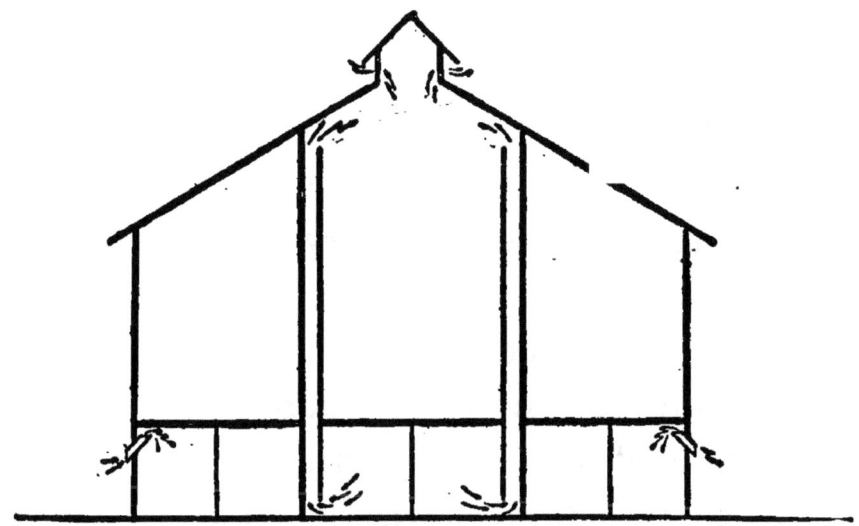

Fig. 4.—Diagram showing ventilating system. The grain chutes may be utilized in this manner

Ventilation of farm buildings is far more important than is commonly believed. Proper ventilation will save you money in building, for if you arrange for a constant and sufficient supply of fresh air, the stock may be crowded as close together as desired. Consequently the building may be smaller for the same number of animals. The best ventilators open into the top of a room, allowing the cold air from outside to sink down through the upper layer of warm air. In this way the warm air at the top of the room is not wasted, as usually is the case, but is utilized in warming the room. The exit for the impure air should be arranged to open close to the floor, for that is where the impure air is, and that is where most of the animals are breathing. If these matters are properly attended to, neither low ceilings nor ground cellars are objectionable if the rooms are dry and light. Pure air and bright sunshine are as necessary as safe water and good food.

For a silo nothing but a round shape should be considered for a moment. For the same cost of construction, it will hold more material and the ensilage will pack better. There will be less waste due to air exposure or freezing on the sides, and the silo will be better able to withstand the enormously heavy strains upon it. The side pressure on the walls amounts to slightly over ten pounds per square foot of wall space for each foot in depth, and the silo must be deep. Most of the early silos were made square

Fig. 5.—The tall, round silo is best. This one is built with concrete blocks

Fig. 6.—An example of true building efficiency. The chute between the silos

or rectangular and shallow. They were not successful until weights were used to press the ensilage tightly into place and expel the air. To-day the best silos are built tall and the weight of the ensilage itself keeps the mass packed tight. To do this the walls must be as smooth as possible. Otherwise the ensilage at the sides will not pack down and will spoil. Many farmers have this trouble, due entirely to rough walls, and the only remedy is to tramp the material down at the sides continually during filling.

For icehouse construction the round shape may be more economical, even if the ice is packed in a square block, provided the corners of the block are brought up close to the sides. A much better plan, however, is to pack the ice into a block approaching in shape the inside shape of the icehouse. When this is done, the saving in building material for the same amount of ice stored will be at least 10 per cent. Nothing makes a better icehouse than an old silo properly repaired and the floor well drained.

There is another general plan which should be considered in every building operation. It is to make every beam and every board do as much service as possible. If there must be heavy uprights to divide the stalls, let them support the next floor. If you must build a partition for a box stall, let it do double service by being also the wall of the harness room. If you have a hay chute or a grain chute from the upper floor, continue it on up to the roof and let it do duty as a ventilator for the lower floors. Plan to make the building 100 per cent. efficient.

CHAPTER II

The Farm Icehouse

The advantage and convenience of having a good supply of ice is far beyond the small cost of gathering it, not only to the general farmer but also to the dairyman, the country merchant, and the rural dweller. Many times a vacant shed, a corner of the barn, an unused cellar, an empty silo, a vegetable storehouse, a dry well, or even an old cistern in the ground, when properly cleaned and fitted up in accordance with the principles here given, will serve as a satisfactory icehouse for many years. The successful icehouse is not necessarily the most expensive one. In southern Virginia a hole dug in the ground entirely above the water level and lined with native clay held ice satisfactorily through the fall. Its only covering was of leaves and pine boughs. This is the type used by the Romans in the early ages to keep snow, and it is now quite common in many parts of this country.

As another extreme of simple construction, a farmer in New York State built four walls of single thickness of board supported by upright green poles freshly cut from the woods. He filled it with ice surrounded by a foot of sawdust, using a layer of sawdust for a floor and another layer for covering. It had no roof, doors, nor windows, and the ice kept all summer without much waste.

It is obvious from these two examples that building material, whether wood, earth, stone, brick, or concrete may not be the deciding factor in the keeping of ice. The secret is in the strict observance of four principles all of which finally reduce to one, namely, good insulation. The four principles are: first, there must be good under-drainage to carry off the melted ice, for otherwise it would form a conductor of heat to the remainder of the ice stored, and would gradually melt it from underneath. Water melts ice much faster than air, for the latter merely affects the surface while the former penetrates throughout. Second, there must be perfect ventilation at the top of the ice in order that the covering of sawdust, straw, hay, moss, or leaves may be kept as dry as possible so that it will not form a conductor for the heat from the air and melt the ice on top. Third, the ice must be packed so as to prevent the circulation of air through the mass, for there is certain to be some heated air enter into the house when the doors, windows, ventilators, or top are opened. These currents of air rapidly warm up, while dead air does not readily become heated because of the fact that air is a very poor conductor of heat. Fourth, good insulation at the sides and bottom must be carefully provided.

The size of the house needed may be determined from the fact that a ton of stored ice occupies approximately 42 cubic feet of space. The average size of house for a small farm is about ten feet high from the ground to eaves with an inside area 12 x 14 feet. After allowing for the space occupied by the sawdust around and under the ice, this will give room for the storage of from 25 to 28 tons of ice. A cubic foot of solid ice weighs close to $57\frac{1}{4}$ pounds, so that 35 cubic feet of solid ice would weigh a ton. From this we can estimate the amount possible to cut from a pond. The thickness of the cakes cut ranges from six inches in the central states to 16 and even 20 inches in the north. Probably 12 to 14 inches is the average. The cakes are cut various sizes also, perhaps 12 x 16 and 16 x 16 are common sizes, but this is not important. Assuming cakes 12 inches thick and 12 x 16 inches, there will be 26 of them to the ton, each one weighing $76\frac{1}{3}$ pounds. In the field, allowing for breakage and waste, a surface of 50 feet square will harvest 45 tons of 12-inch ice.

Having determined upon the size of house and the outlay of money that can be afforded, it remains to determine the material to be used and the plan to be followed. Beyond any reasonable doubt wood is better in many ways than stone, brick, or concrete for icehouse construction, although any of these may be used with satisfaction if the ice is packed far from the walls and well insulated from them by ten or twelve inches of sawdust. The only objection to wood which any one can have is its tendency to rot under the continued influence of moisture inside and dryness outside. For this reason cypress is to be highly recommended as a serviceable wood, although pine will last for some years and is quite generally used in practice.

For a foundation concrete is best, all things considered. Let it go into the ground below the frost line and extend a foot above ground to keep the sills dry. Unless the soil is well drained, there should be a main ditch with side branches cut in the floor, covering the whole space below the ice, the main ditch leading out on the lower side. Fill the ditches with broken stone, crockery, brick, or clinkers, and spread a thin layer over the whole floor. On top of the stone place a layer of straw covered with a thickness of coal ashes. On top of the ashes floor boards may be placed with cracks between them to allow free drainage of the water from the melted ice. More often, however, the boards are dispensed with and an eight or ten inch layer of sawdust put directly on the ashes, the ice being packed on that.

The walls may be either single or double, but should be built with matched boards or papered with tar roofing paper. I should recommend both. The paper is cheap, costing $1.50 to $2 for a 500-foot roll, so it does not add much to the cost of the work, but it does give a much better house. If the single walls are papered it should be done on the outside, of course, while if the building is made with double walls, the papering should be on the sides within the air space. Double walls are much better for insulation and may be easily provided by nailing the boards on both sides of the 2 x 4 joists used as uprights. This leaves a four-inch dead air space between the walls which should not be filled with sawdust nor with anything else. The best insulator we have is dead air, and the purpose of sawdust, felt, wool, shavings, and such substances is merely to keep the air dead —that is, these substances prevent circulation of air by catching small quantities in the spaces between the particles. The use of these substances is not to be recommended either in icehouses between walls or in the walls of cold-storage boxes. In either case the filling would become damp and remain so, thus rotting the construction from the inside. In cold-storage boxes it also will absorb and retain the odours, making the box unfit for keeping eatable produce. Furthermore, when damp such fillings are reasonably good heat conductors.

In the air space between the boards, in the icehouse construction, every three or four feet up there should be a strip of tarred paper tacked to form a horizontal partition, thus preventing any up and down circulation of the air. The result of this construction is that the ice is surrounded by walls consisting of a large number of boxes containing dead air. These boxes will be from three to four feet square and four inches thick (the thickness of the air space).

The sills of the house should be laid directly on the concrete foundation and in close union with the concrete to prevent entrance of the air between them. In my experience it has been found well to lay a coating of tar or asphalt on the foundation walls and on this put the sills, thus making an air-tight job. There must be no entrance of air underneath the ice. It is true that a small amount will enter through the drain if the latter is not trapped, but this is not sufficient to do any harm. In a commercial house of large size, however, the drain should be of tile and trapped as it comes from under the icehouse. Preferably, too, there is a drain around the foundation on the outside, both of the drains being brought together and led away to a lower level.

The roof for a small building may be almost anything to shed the rain, keep off the sun, and provide good ventilation. The latter feature is the one most important point in connection with building the house. The ventilators should be closeable and kept closed on foggy days and nights. For this reason trap-doors on the sides and roof are preferable. The roof thould be a V-shaped or hipped roof, with trap-

Fig. 7.—Simple roof ventilators for icehouse construction. This view shows a trap-door arrangement for the end walls, giving opportunity for proper ventilation

doors at each end and at the ridge. Near the top of each end wall arrange a small door. Each fine, dry day open one of these doors and the opposite trap so that the air may circulate freely and keep the top dressing or covering of sawdust dry. This top dressing should not be too thick, the practice being to have it from eight to twelve inches. The dressing must be looked after and kept dry at any cost. It will be found helpful, although a nuisance, to divide the top layer by a thick layer of newspaper.

In packing, the first layer is commonly placed on edge rather than being laid flat. There is no less wasting that way, for, although each cake wastes less, there are more cakes on the floor. Sometimes this plan is followed throughout, the advantage being that in breaking the ice out there is less adhering surface between the cakes. It is harder to pack this way, however, and the liability to undue sidewall pressure is greater. At least every third layer, no matter how packed, should be laid so as to break the joints of the previous layer that there may be no circulation through the mass. The packing can be done up to within six inches of the side walls if a double wall is used, and up to within eight or ten inches if a single wood side. As stated before, if concrete, stone, or brick is used, there should be from ten to twelve inches left around the sides. In every case the space left should be filled with sawdust lightly tamped into place but not rammed tightly. Hard tamping forces the sawdust down so solidly as to remove most of the air, while light tamping keeps the mass porous but yet held together tightly enough to retain the air and prevent its escape or circulation.

In conclusion, it should be said that the cakes must be cut as true as possible, and no small pieces or broken cakes should be allowed to enter the house. The ice should be packed in freezing weather so that the cakes will be dry and not freeze together in the house. Each cake should be kept an inch or an inch and a half from its neighbour on every side.

CHAPTER III

The Principles of Cold Storage

Most progressive farmers have learned the value of the individual icehouse, yet have not realized that the most economical way of using the ice cannot be developed without a properly constructed cold-storage chamber. Creamery and coöperative cold-storage chambers are getting to be quite common now, and their importance is realized. As the farmer observes them in use he will undoubtedly come to appreciate the value to him of a similar house built on a smaller scale.

The details of construction may, as in the case of the icehouse, be widely varied to suit particular needs. There are certain fundamental principles which can be laid down for guidance, however, and close adherence to them will mean success in the construction. Satisfactory insulation can only be obtained through the use of double walls for the chamber, in this way providing a dead air space between the walls, as that is the best form of protection. The air within the space must be *dead* air so the walls *must* be airtight to give satisfaction. There are many other ways of insulating, as by filling the space between walls with some so-called "non-conducting" substances such as the following named in the order of their desirability: hair felt, slag wool, wood ashes, chopped straw, charcoal, cork, and others. The insulating properties of these substances are largely owing to the fact that they enclose in the tiny spaces between the individual particles small amounts of dead air which cannot escape. That air is the insulator. For this reason the substances cannot be packed solid, and should be lightly tamped into place rather than rammed hard. For cold-storage work it should be borne in mind that something must be chosen which does not readily absorb moisture and odours. There is no one substance which does not do this to some extent. If the building can be built with matched boards and the dead air space lined with tarred paper, the space need not be filled with anything. In fact, a filling would be a decided detriment.

Moisture has the property of absorbing many gases and impurities from the stores, so it is very desirable that the air in the chamber be kept as dry as possible and that the moisture which it does take up be removed. In this way the air may be purified. The way in which it is accomplished is by providing proper circulation of the air in the storage chamber and thus cooling the stores by circulation of the cold air in contact with them rather than by radiation. Unless cooling is done in this way the moisture which the air contains will be deposited on the stores and not on the ice. This, of course, will cause some of the packed material to become tainted.

To get a good circulation it is necessary to appreciate the fact that cold air drops and warm air rises. All that needs to be looked out for then is to have the ice box above the level of the storage space floor and to introduce the cold air at the bottom of the storage space, providing an outlet and return at the top of the chamber for the heated air to go back to be cooled and deprived of its moisture. For a small chamber it will be satisfactory if the cold air is allowed to enter all along the lower edge and the warm air taken out the upper and diagonally opposite edge. This will make it necessary for the air to cross and circulate all through the storage space before reaching the outlet. In a larger chamber the cold air could be introduced at the centre of the floor and taken out at each of the upper side edges. In a still larger room the cold air may be introduced along two side edges at the bottom and allowed to go out through two side edges at the top. Shields or deflectors, which may be made of wood painted with enamel, should be placed so as to prevent the cold air as it warms up going from the inlet opening directly to the outlet opening without circulating through the room. These deflectors should slope from the bottom up and be placed just over the cold-air inlets so that as the cold air warms it will rise along the deflector toward the outlet. Care must be taken not to place the deflectors so as to pocket any warm air—that is, do not make them so that any body of warm air will be caught in an upper corner and have to go downward to escape. Deflectors are only necessary where the outlet is nearly over the inlet and a path from one to the other does not lead through or near the centre of the storage space.

Ventilation is essential, but, except in very large rooms, it is satisfactorily taken care of by the opening and closing of the entrance door.

The packing of stores in cold storage is a science in itself and can only be taught by experience. The general rule is of value, however, and will take care of most difficulties. It is to pack the stores fairly close together and leave a space between them and the walls so as to allow a path for the circulating air. Never pack up close to the walls.

CHAPTER IV

THE WATERPROOFING OF CONCRETE

CONCRETE needs no waterproofing if it is properly mixed and laid. Water leaks through because the mass is porous. If we consider the materials entering into concrete construction and the theory upon which the structure is based this fact will become clear to us. Concrete contains cement, sand, and stone. The stone, if used alone, is extremely porous, for the spaces between the individual stones are quite marked. The theory is that the sand used goes to fill these spaces. Yet even then there are spaces between the sand grains and water will pass quite readily. These spaces, however, are filled with the cement, the particles of which are so very much smaller than the grains of sand. The cement particles do more than merely fill the spaces between the sand grains. They cover the individual grains and cement them together, embedding the stones within the whole mass.

It is apparent that if all the spaces are filled water cannot leak through, while if the mass is filled with tiny pores not only will water pass through but these pores or tubes will suck up or absorb water from the ground and from the moisture which condenses from the atmosphere. Such will be the case if the concrete is "poor" or "lean"—that is, if it does not contain the proper proportions of materials or the proper sizes of particles to enable the cement to thoroughly unite the ingredients. Cement is the costly part of the concrete and the temptation is to use as little of it as possible. This does not pay in building any foundation walls, cisterns, tanks, and such structures where it is necessary to prevent the flow of water through the walls. If the wall does leak, there are but two things to do in order to remedy the defect: Either the pores must be plugged up with some substance which is not porous to water, which is not dissolved by water, which may be easily and cheaply applied, and which will not chemically attack the concrete, or a separate layer of waterproof material must be laid against the surface of the concrete, using the concrete merely for its mechanical strength and trusting entirely to this auxiliary layer to repel water.

It is perhaps obvious that in every case where it is possible to do so the waterproofing materials or layers should be applied to the concrete on the side next to the water. Unless this is done, the concrete will always contain water and the waterproofing will simply prevent the water from flowing out. Under these conditions neither the waterproofing nor the concrete is apt to give entirely satisfactory service. The construction of waterproof concrete needs carefulness and thorough workmanship, but when we consider the difficulty of making a real, lasting job of waterproofing, after a wall has commenced to leak, it will be seen that care in the mixing and laying is more than repaid. There are several good waterproofing proportions differing but slightly. The 1-2-4 mixture is most commonly used. This means one part of cement, two parts of sand, and four parts of gravel or broken stone. With these proportions, one bag of cement mixed with the proper amounts of sand and gravel will give a bulk of finished concrete measuring about four cubic feet.

Portland cement should be used for all work of this kind. It may be purchased ready for use in either bags or barrels, but the bags are far more convenient for handling. The sand and stone may be obtained anywhere. It is important, however, to have them clean, with no mud or sediment clinging to them or mixed with them. To be sure of this they may be piled on a sloping board platform and thoroughly drenched with water, turning them over several times in order to clean the bottom and interior layers. The sand must be coarse or a mixture of coarse and fine for the most economical results. The total spaces between the particles of fine sand are more and the total surface of the sand particles which the cement must coat is greater with fine sand. Hence, the finer the the sand the more cement must be used and the more expensive the concrete. Coarse sand, with a small amount of fine sand mixed in, is desirable, for the fine sand fills up some of the spaces between the coarse particles and makes a more solid concrete. It will always pay to buy coarse sand rather than use fine sand which is free. The appreciable saving in concrete will be great.

Contrary to the prevalent idea, gravel makes a better concrete than broken stone. It is more dense and it is stronger after it has aged. Particularly is this true of a gravel of quartz pebbles.

The concrete should be mixed a little wetter than is ordinarily done, and the mixing must be thorough in order that the proportions may be properly intermingled. In laying, great care must be exercised not to separate out the ingredients by pouring or dropping from a bucket or barrow through a considerable height. If this is done, the job will be spoiled. After laying, the concrete should be tamped slightly in order to drive out the air and fill the voids or holes. Following this, the surface layers should be spaded. That is, a spade is placed in between the wall and the form and drawn up and down in order to slightly "puddle" the surface, driving back the gravel a little and leaving the surface with a grout as nearly airless and non-porous as possible.

By following the suggestions given, the concrete cannot be penetrated by water, but concrete that will not absorb moisture to some extent cannot be made. It is only possible to prevent absorption by adding some waterproofing compound to the concrete when mixing, or by treating the surface of the concrete after it is laid. The mixture laid under the above conditions is dense and close grained due to the excess of cement, and it is without air bubbles because of the excess water. It is filled with very tiny capillary tubes which will not allow the passage of water yet will absorb it in small quantities. This is undesirable in many places where concrete is used, and to prevent it some one of the following methods are employed.

If it is old work which is to be protected, only surface coatings can be used, and their object is a filling of the pores spoken about. Four substances are commonly used for this, namely: neat cement, asphalt, paraffin, and an alum-soap compound. This last is known as the Sylvester treatment, and is one of the most effective. In a different form it is used also for new work as will be explained later. For surface coating a hot castile soap solution is made by dissolving three quarters of a pound of the soap in one gallon of hot water. An alum solution, of one half a pound of alum to four gallons of water, is then prepared. The substances are thoroughly dissolved and alternately applied to the wall, the latter being perfectly dry. The hot soap solution is first applied, a flat brush being used and care being taken to avoid bubbles covering the work. After this coat dries for twenty-four hours, a coating of the alum water is put on and allowed to dry for a similar length of time. In this way, alternate coatings to the extent desired may be used, allowing a full day to elapse between the coatings. There is a chemical process which takes place between the substances used, the resulting compound plugging up the pores in the cement. The cost of this process for two coatings of each material will be from 35 to 40 cents per square yard.

Paraffin, although rather expensive, is often used for small jobs. It may be melted and applied while hot, the walls also being slightly warmed, or it may be dissolved in some solvent such as benzol, xylol, or even benzine of the common kind, these liquids quickly evaporating. Several coatings will be needed, and each coating will cost in the neighbourhood of 50 cents per square yard. If you do the work yourself and do not count the cost of your own time and labour, this cost will be materially reduced.

Asphalt and other bituminous products are the easiest to handle and the surest of results in unskilled hands. They are applied as liquids, allowed to dry, and further coatings given. Probably the cost for two coats will not exceed 25 cents per square yard.

Cement grout is a mixture of cement, sand, and water or just cement and water, very liquid and applied like paint. It is not very efficient when used on old concrete, for it readily peels or cakes off after a short time. For a temporary repair this or a mixture of the same substances just plastic enough to handle with a trowel is the most universally used.

The surface coatings spoken of are as valuable for concrete blocks, brickwork, and porous stone as for straight concrete work. Good brick needs very little attention, although it will absorb from 3 to 5 per cent. of its weight of water, but such brick is expensive and seldom met with on the farm. The common brick used will often absorb from 15 to 25 per cent. of its weight in water. Concrete blocks, especially if made by the continually tamping process known as the dry process, are extremely porous.

While the above coatings appear to be satisfactory for simple work, in large structures such as dams, reservoirs, and sewers much more care must be taken. Strong layers are used because of the heavy water pressure against them. Felt or burlap saturated with tar or pitch, rolled in a continuous layer against the wall and held there, is not only a satisfactory water retainer but also prevents the leakage of foul gases which chemically attack the concrete. A method known as the integral process is practised where it would be too expensive to use the thorough workmanship described in the early part of this article. This consists in the addition to the cement, when mixed, of some fine, dry powder consisting of extremely small particles, usually alum and lime. These, because of their size, may fill in the spaces between the cement and sand grains and make the whole structure more dense. Usually only the cement which lies near the surface is thus treated. Still another treatment is to add some soap or oil emulsion to the mixture. This forms a jelly within the concrete and fills the pores.

Lastly, the well-known Sylvester process before mentioned is used. Alum is added to the cement and castile soap is added to the water with which the mixture is made. Chemical action then goes on in the mass, forming a compound which, as before, fills the spaces. Many, many other substances may be used. In fact, one farmer in waterproofing a cracked wall filled the cracks with corn stalk pith and wet it, causing it to swell and fill the cracks completely. The whole object of waterproofing is to fill all holes, pores, and cracks. Any method of doing this satisfactorily is entitled to consideration.

CHAPTER V

ARTIFICIAL STONES AND COMPOSITION FLOORING

THERE are a number of artificial stones on the market which may be readily used by any one. They may be used in blocks, as concrete blocks are used, or as a covering for making a structure either fireproof or waterproof. The latter is perhaps of most interest to farmers. Often it is desired to cover old floors or, with the installation of bathroom fixtures in the house, it is desirable to make a waterproof floor. For this latter purpose the various modifications of Sorel stone are highly recommended. Its strength and hardness exceed that of any other yet produced, and it is one of the cheapest of the artificial stones. For stable and stall floors it is also of considerable value, for it is sanitary, easy to clean, and wears well.

There are almost as many different varieties as there are users of the stone, for every one makes some little change in the details of mixing. The fundamental thing is to mix in with the various filling substances an "oxychloride binder" which is nothing more nor less than a solution of magnesium oxides and chlorides. This is used to moisten the filling substances in the same way as water is used in concrete work. The fillers here may be almost anything— shredded wood or cloth, sawdust, asbestos, sand, ashes, pebbles, etc. You may buy the material all ready for use from any large paint shop or hardware store and do the work yourself, or you can get any of the companies selling the substances to do the work for you. You may, if you like, mix up the ingredients and make your own stone.

If you buy the material ready to use, you will get two packages. With one kind the packages contain powders which must be mixed together and water then added. With the other kinds of composition one powder and one liquid is purchased and the two mixed. The mixture is made somewhat stiffer or thicker than ordinary cement and is spread on the old floor, or on the flooring built to receive it, to a depth of half an inch. The surfacing must be well done and not left rough. It will "set" overnight and will be hard enough then to walk on if care is observed, but the floor should not be used for three or four days. Probably several months will elapse before the floor reaches an even colour all over. From time to time it will be necessary to remove the white blotches by simply washing, the spots being caused by the chemical action going on in the floor material. After a time the floor will be stone hard and, of course, will be fireproof and waterproof. This is a valuable characteristic, for almost all artificial stone in common with brick and concrete is very porous and open to the absorption of water. Composition floorings, however, are not open to this serious objection. Any desired colour may be added to the composition, the earth colours giving the best results.

Although this flooring has but just been receiving the attention of private builders, it is not a new thing. There are hundreds of patents for different mixtures, and one kind or another has been used on the floors of railroad cars, in public buildings, and similar places for at least twenty years. Recently some of the important patents expired, and during the past ten years as many as fifty companies manufacturing composition have come into existence.

Specifically, the ingredients of one of the best compositions are as follows:

50 parts (by weight) of calcined (burned) magnesite.
15 parts of dolomite (marble dust).
5 parts asbestos (shredded).
15 parts sawdust.
2½ parts silicate of magnesium.
11 parts of earth colours.

Mix the above powder very thoroughly and add the following liquid until the proper consistency is obtained. Frequently, to make a better union of the elements, the above powder is added to 1½ parts of muriate of ammonia.

The liquid: equal parts of water and chloride of magnesia solution.

In another composition flooring the materials are mixed at the shipping point and the receiver adds water and burned or calcined magnesite. In this the specific materials are:

85 parts magnesium chloride solution.
36 parts of any filler such as sawdust, ashes, etc.
25 parts of infusorial earth or fossil flour.

Add to the above:

100 parts of pulverized burnt magnesite.
43½ parts of water.
Desired colouring material as red oxide ochre, etc.

All of these substances are cheap. The mixtures as retailed by manufacturers are about fifteen cents per square foot of floor surface for the substance and an equal amount for doing the work. Unless you are willing to take great pains with the laying and finishing of the floor, an expert should be allowed to do it.

By using the above mixture but substituting large pebbles or stones for part of the filling material suggested, a first-class concrete is obtained. The same mixture, also omitting the filling material and colouring matter, and adding the proper sand or sharp, small stones, may be used for the formation of grindstones, emery wheels, etc.

Another common artificial stone used for blocks and slabs is known by the name of Ransome stone. It is formed by mixing sand and the silicate of soda in the proportions of a bushel of sand to a gallon of the silicate. The mixture is now very easily worked and is rammed into molds for blocks or ornamental shapes, or may even be rolled into slabs for walks, paths, and such purposes. The slabs or blocks may then be cut in any desired way. After being made in just the shape and size desired, they are immersed in a hot solution of calcium chloride which is under pressure so as to force it through the pores. The chemical action between the sand, the silicate of soda, and the calcium chloride forms a hard and non-porous cementing substance in the spaces between the sand particles, while some sodium chloride is formed in the process and must be washed out by thoroughly drenching the blocks or slabs with cold water. The sand enters into the action but slightly, and any gravel or broken stone of small size may be mixed in.

The "Beton-Coignet" artificial stone is still commonly used because of its quick setting property, its strength, and its ease of manufacture. The ingredients in the proportions of

    4 parts lime,
    1 to 2 parts hydraulic cement,
    20 parts sand

are very thoroughly mixed by hand. In this great care must be taken to insure thorough mixing. The mass is then again mixed in a mixing mill of any kind with a very, very little clean water, just enough to moisten the substances slightly. Molds can then be rammed full of the mixture just as is the practice with concrete. The finished stone occupies slightly more than one half the bulk of the dry mixture, and weighs practically the same amount as Portland cement concrete—that is, 140 pounds per cubic foot.

CHAPTER VI

PAINTS AND PAINTING

THE coming of spring should be a signal for painting everything that needs it, whether house, barn, fence, or machinery. Particularly should machinery be looked after, and emphasis cannot be too strongly laid on this point, for few things are so neglected as machinery on the ordinary farm. Not all paints are of equal value for these different jobs. What is good for iron is not good for concrete, and the paint which is so satisfactory on the house may be of little value for the wagons. A paint for woodwork consists of some dry material for colouring, a lead or a zinc base, a drier, and a vehicle or liquid. It is the vehicle which is often wrongly chosen, and in some ready-mixed paints the vehicle is the part which is most likely to be adulterated. For outdoor work, except decorations, boiled oil is considered to be the best. For indoor work linseed oil and turpentine are preferably used. A little drier, litharge for dark paints and sugar of lead for light paints, should be added to each batch of paint mixed.

Undoubtedly linseed oil paints are more expensive than others, but they are well worth the difference in price. This oil enables the paint to spread well, dry hard and opaque, and leave a protecting skin over the wood surface. If adulteration is practised with resin oils, mineral oils, or fish oils, the paint will either remain sticky forever or will harden quickly only to soften again in a week or ten days. Particularly should dark-coloured paints be looked upon with suspicion unless purchased from a thoroughly reliable dealer, because such paints when cheap usually contain only unrefined resin oils which soften up within two weeks of the first drying. They never harden again but give constant trouble.

One of the best paints for roofs and machinery is a mixture of red lead and linseed oil. Another good metal paint is known as asphaltum varnish. It may be purchased ready for use, and when applied leaves a splendid wearing, shiny black surface which thoroughly protects the metal.

For painting jobs requiring the covering of a large surface, the paint may usually be sprayed with much less labour than if the application is made with a brush. Almost any paint may be so applied if it is made thin enough. Use the ordinary spraying apparatus which is used for disinfecting and orchard spraying. It may be readily cleaned and will suffer no injury by such use. Probably whitewash is more often applied in this way than any other paint. Particularly for fences and outbuildings this method means a great saving in time. Yet ordinary whitewash is not as economical as cement whitewash. While the former requires frequent renewals, the cement wash often remains satisfactory following several years' wear. The combination is best made in the following proportions: Mix together one peck of white lime, a peck and one half of hydraulic cement, six pounds of umber, and four of ochre. The lime is first slaked and mixed with two ounces of lampblack moistened with vinegar. Then add the other ingredients. Allow the paint to stand for three hours or longer, stirring occasionally. The addition of half a pound of Venetian red renders the appearance more pleasing and adds to the value of the paint. If ordinary whitewash is used at all the addition of a little glue or a small amount of flour mixed with boiling water and poured in while hot will prevent the whitewash rubbing off so readily.

For finished interior work, varnishes are best to use. They give an extremely hard surface, which protects the wood beneath, and they are easy to clean thoroughly. It is not advisable for any one but an expert to attempt to mix them at home, for many good ones are on the market, as well as many worthless mixtures called varnishes. True varnish is a solution of resins or gums in some suitable liquid such as alcohol or oil of turpentine mixed with linseed oil. Those in which alcohol acts as the solvent are spirit varnishes and are inferior in many ways to the oil varnishes, chiefly because the alcohol evaporates entirely, leaving the varnish so hard as to easily crack and chip. The oil varnishes, on the other hand, should never get brittle.

CHAPTER VII

Lightning Rods and Rodding

EVAPORATION which takes place at all points of the earth's surface is believed to cause electrification of the particles of moisture in the atmosphere. As these particles unite to form clouds, the clouds become charged with electricity. The potential of each cloud rises higher and higher and the earth beneath becomes charged by influence, or, as scientists say, by induction. This induced charge on the earth is of opposite sign from the charge on the cloud. Presently the difference in potential between the cloud and the earth becomes so great that the air between them breaks down and a passage of electricity takes place. This is the lightning spark. This spark discharges only the electricity accumulated on the under surface of the cloud, and when that discharge takes place the cloud must adjust itself again, and it does so by discharges between the parts of the cloud so that there is much internal action, which accounts for the apparent boiling of the upper part of the cloud. When the cloud is readjusted, further sparking can take place from the same under surface, which explains why many lightning discharges take place during the same storm.

Sometimes the cloud, in place of discharging to the earth, discharges to another cloud. If that other cloud is of small capacity it may overflow and discharge to the earth. These charges are often disastrous for reasons given later.

Now, if there were a conductor, such as a metal rod, extending from the cloud to the earth, the charges would be equallized, without a lightning spark, by a passage of the electricity over the rod. As there is no such conductor, the spark chooses the easiest path to follow—the line of least resistance. That accounts for its jagged appearance, as the easiest path may not always be the straightest path. Dust particles, a current of moist air, a current of hot air, or a draft is very likely to be followed, as such are better conductors than cold, clean dry air.

The protection of barns or other buildings

Fig 8.—Showing the discharge between clouds and the overflow to earth

from lightning involves, then, providing an easy path for the lightning to follow to the ground, for it must reach the ground and will choose its own way, however disastrous, unless we choose for it.

When the earth is charged beneath a charged cloud, the buildings are charged, too, and being nearer to the cloud are apt to be struck unless the charge is dissipated. It has been found that if an electrically charged body be connected to a metal point, the charge rapidly leaks off the point. This, then, is the second function of a lightning rod—to dissipate the accumulated charge on a building, and thus prevent it from being struck. This cannot be done, however, in the case of overflow charges as described above, because the overflow takes place so suddenly. Hence, those strokes are particularly dangerous.

From the standpoint of lightning protection, then, if the barn doors and windows are left open, there is a great draft which may offer a path to the lightning discharge, and there can be no adequate protection from lightning at the doors and windows if they are open. There is, too, considerable heated air and some dust passing out which offer an easy path for the lightning, and a lightning charge passing inside the barn is sure to set the hay on fire. On the other hand, if the doors and windows are shut and ventilation provided at the top for the steam to escape, there may be two crossed

Fig. 9.—Lightning protection for the roof ventilator. The wire should make no sharp bends

arches of metal over the opening with a sharp metal point at their joint and connected by a direct line to ground, thus affording reasonably good lightning protection should the warm air act as a conductor for the lightning stroke. Moreover, the upward flow of moist, warm air over the point would help greatly to cause any charge accumulated in the barn to leak off the point, if the whole system of protection was connected to the point. Then, if the barn is well rodded, and the ventilating opening properly protected, there is not so much danger that the hay will be set on fire even if lightning strikes the barn, as it will reach the ground probably without going inside.

Absolute security from lightning can be obtained only by a large outlay of money. If the building to be protected is well insured, and a fire would mean merely the loss of the building but no loss of life, it is not a business proposition to expend much money for additional protection in the form of lightning rods. On the other hand, if the building is not heavily insured or if a fire would be disastrous, it is a paying investment to rod the building well.

The only way to completely protect a building is to enclose it wholly in metal and carefully connect the metal to the ground. This is usually too expensive, so that as a compromise a building is enclosed in a network of wire and the latter well grounded. This is the plan used on the

White House and on the Washington Monument, the only important government buildings rodded at all. It is the practice followed abroad extensively, and has been recommended many times by leading experts and engineers.

There are several other fundamental considerations. The metal of the rod, the joints between the parts, the nature of the ground connection, the fastening of the rod to the building, and the construction of the discharge points all require careful thought and workmanship.

As to the choice of metals, prominent scientists differ in theory, although all agree on practical details. There is no question but that copper is a better conductor of electricity ordinarily than iron or any other common metal. Yet a discharge of lightning differs from the ordinary passage of electricity in so many respects that the metal which is best in ordinary use may not be best as a lightning rod. As long as the metal is in first class condition, the joints perfect, and the whole thing well protected from the weather so that it won't oxidize or rust, any metal will give satisfactory service. The chief advantage of copper is that it will stand the weather better than iron, but the latter when galvanized and painted will last a long time and is cheap.

If copper is used, the stranded or braided cable is cheaper, lighter, and better than the solid wire or rod. If this cable is obtained hollow, there is a still greater saving, for the interior of lightning conductors is not used in the passage of electricity but is worthless except to give mechanical strength.

Whatever metal is chosen, it must be kept in first class repair at all times, or it will be of no value. Abroad many of the governments use two strand, galvanized, barbed iron wire entirely, even for important work. The barbs are kept sharp and the wire kept free from rust. With these precautions such a system will give satisfaction.

The chief use of a lightning rod is to prevent a stroke of lightning taking place rather than in conducting a discharge to earth, although the latter must be provided for. The barbed wire is particularly desirable because of its multiplicity of sharp points, whereby what is known as the "silent discharge" can take place from a building. This prevents the gradual accumulation of an electrical charge, and no lightning stroke can take place. At suitable intervals on large buildings larger sharp points, say six to eight inches, should be placed in a vertical position and well soldered to the main rodding. The large points should in every case be as nearly vertical as it is possible to get them.

For a ground connection, it is not sufficient to stick the end of the rod in the ground. Such carelessness might prove disastrous. The ground is the vital part of the whole rodding system, and too much care cannot be given to it. The grounding device should be buried at least ten feet deep in moist earth, and should be perfectly connected to the main rod by welding or soldering. It should be thoroughly protected from rust or other deterioration, and care should be taken that the earth is closely packed around the rod where it enters the ground. The best grounding arrangement is a large piece of metal or a very large bundle of wire, particularly barbed wire.

To show just how the rules laid down above should be applied, we may take the case of a barn for example. Assume that an inexpensive system is desired, and so barbed wire is to be used. First, lay a double strand along the ridge pole from the back peak to the forward peak, then down the sloping edge of the roof to the eaves, along the eaves, up the sloping edge at the back end to the peak, down on the other side and along the opposite eaves, up the remain-

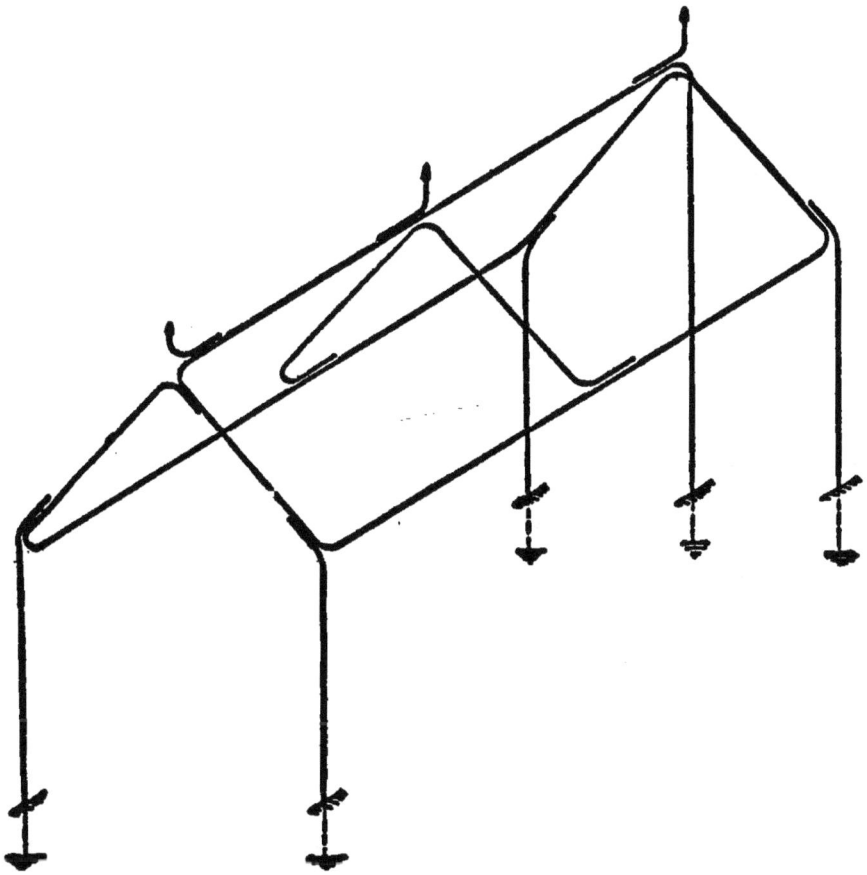

Fig. 10.—The method of bending barbed wire to form an enclosing network for lightning protection

ing sloping edge to the front peak where we started from. Here we may cut the wire, leaving a length of four or five inches, which should be tightly bound to the first wire with copper wire. This joint should be flooded with solder. At the back eaves sufficient wire must be left to reach down to the grounding device. Where this ground wire crosses the other the two should be bound together and soldered.

About every eight feet along the ridge a cross wire is placed, extending down to the eaves wire on each side, the joints all being bound and soldered. If the barn has a gabled roof, another wire should extend along the outer ridges, being carefully connected to every cross wire. All the wires must be fastened directly to the wood by means of double pointed staples. Under no circumstances should insulators be used, as they render the whole system useless. Moreover, all metal on or near the barn must be connected to the lightning rod. Any wire fences nearby must be connected to the ground wires. It is well, also, to thoroughly connect all wire fences on the farm to the ground at intervals of fifty feet, as by so doing stock standing near the fence in a thunderstorm will not be in danger.

The ground wires for the barn should extend from all of the lower eaves corners and from the back peak directly to the ground in as straight a line as possible. They should hug the woodwork closely, but not follow all of the bends and corners. If the door is on the long side, a ground wire should extend also from the front peak. Each ground wire must be bound to the top network and soldered or welded.

For each grounding device, coil up a hundred feet of the barbed wire in a ball and bury it ten feet deep. This can be a continuation of the ground wire. In covering the ball add water to the dirt as it is thrown back in the hole, and it can be stamped down much tighter around the grounding device than if dry earth is used.

At both peaks and about every twenty-five feet along the ridge erect sharp points six or eight inches long. Preferably they are made of heavy copper wire filed to a point at one end.

The bottom end may be bent for binding and soldering to the wire on the ridge pole. Similar points should be placed along all the ridges on a gable roof.

If the work is properly and carefully done, the result will be a wire cage solidly joined throughout and completely covering the barn. The wire will have a multitude of sharp points and will be thoroughly connected to the ground in several places. In the case of a very long barn there should be extra ground wires from the eaves down at the middle points of the long sides. The whole, except the copper points, may be well painted frequently.

PART II

FARM WATER SUPPLY AND SEWAGE DISPOSAL

The Sources of a Pure Water Supply.
Running Water for Fifteen Dollars.
A Sand Filter for Rain or Brook Water.
Softening Hard Water.
The Hydraulic Ram and the Ram-pump.
Disposal of House Sewage.

CHAPTER VIII

The Sources of a Pure Water Supply

To have a real knowledge of the conditions likely to affect water purity the farmer must know the essential features of good wells and springs and how to protect them from contamination. It is with the hope of acquainting him with these problems and their solution that this chapter is written.

The distance to the water table or water level determines how deep a well must be dug into the soil, for to be successful it must go below the level of the water table. Then the water will find it easier to flow into the well from the soil immediately around the opening than to continue to seep on to the impervious layer. This causes a lowering of the water level around the well, or a cone of depression as it is called. The water still farther out flows sideways into the depression and on into the well. Finally, the level of the water in the well is the same as that in the surrounding water table. If, now, some of the water is pumped out of the well, its level lowers and water flows in from the soil around it. Obviously, then, the deeper the well is below the water table the greater its capacity will be and the faster it will fill up as water is pumped away, because the head of water causing the flow into the well is the distance the surface of the well water is below the level of the water table.

Rain water as it falls from the clouds is as pure as can be after the first fall has washed the impurities out of the air. When it strikes the ground it becomes contaminated with the various impurities on the surface. As it sinks through the soil, however, these impurities are taken out of the water partly by the bacteria which are so plentiful in the upper soil layers, partly by filtration or straining, and partly by the chemical action of substances in the soil. In order that the purification shall take place, the water must sink through a considerable amount of soil, at least fifteen to twenty feet. A dug well should be so constructed that only water which has been so purified can enter it. This means not only that the well curb must be something more than ten inches above the surrounding soil to keep out the surface flow, toads, rats, and other foreign matter, but the well must be lined in a watertight manner, as by concreting or with stone or brick set in mortar to a depth of from fifteen to twenty feet below the surface. This lining should be pressed tightly against the earth so that water cannot under any circumstances get into the

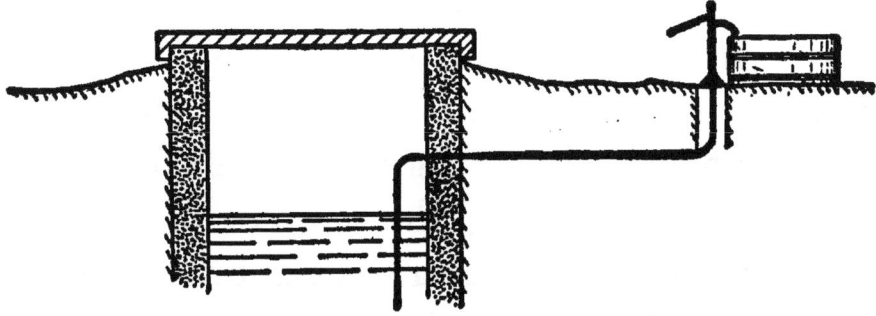

Fig. 11.—The proper arrangement for the top of a dug well. Curb high, cover tight, ground sloping away from well, trough at one side

well unless it has filtered through the soil to the depth of the bottom of the waterproof wall. The top of the well should be closed with a waterproof cover so that no drippings or splashings can run in. The horse trough should not be at the top but should be a few yards to one side so that the drainings may not work themselves directly into the well water. The location of outbuildings, manure piles, pig stys, cattle runs, etc., should be as far away from the well as possible, and greatly below the well if possible. It is much more important to consider the matter from the standpoint of health than from that of personal convenience. The direction of the flow of the underground water to the well cannot be altogether determined by the contour of the surface of the land. Moreover, long-continued pumping at the well may drain the water into the well from a greater distance than would otherwise be the case. Hence, no precaution should be more carefully observed than that of getting the sources of possible contamination as far away as can be. The fact that the well is above the privy does not mean that the drainage from the latter cannot reach the well. It is distance that counts, because that can be depended upon while a point apparently below the well may, when underground flow is considered, be above and draining directly into the well.

It is apparent that a dug well is not always dependable unless certain precautions are taken. On the other hand, deep wells are quite likely to yield water of perfect purity, as far as harmful ingredients go. The reason for this is that a deep well, such as drilled or driven well of several hundred feet, reaches a water-containing layer which lies between two nonporous layers of soil. Somewhere, probably miles and miles away, the water has entered this channel and has filtered through the soil for that great distance, losing all of its bad contents.

An artesian well is of the same nature, but in this case the well opening is made at a point so far below the point of entry of the water in this channel that the head of water is sufficient to force water to the surface. In almost all deep wells the water rises in the pipe above the layer in which it flows. The main precaution to take with deep wells is to be sure the well casing is watertight so that no subsoil or surface water can by any means get into the well.

Springs, as a rule, furnish pure water unless the immediate vicinity gives cause for contamination. The reason is, as stated previously, that the water has flowed for some distance through the soil. The exception is a spring in a rocky district where the water has not sunk through any great amount of soil, but has merely percolated through and over the rocks, finally coming to light at a convenient point. Such spring water should be used with care and only after strict examination of the surroundings of the stream course. If a spring is used as a water supply source, surface drainage in the vicinity should be made good to prevent any surface waters from manured land reaching the water course. The best treatment is to house the spring in, leading a trough from the spring to a sunken storage-basin which is watertight and tightly covered. Then, from a few inches above the bottom of this cistern, lead the supply pipe to the house or barn.

Rain water is, as mentioned above, a pure form of water except in so far as it is contaminated by the atmosphere. Because of this, in some districts it is quite generally gathered and stored in cisterns for household use. If properly handled, there is slight danger of harmful impurities entering the cistern, yet, on the whole, it is not so desirable as ground water, particularly at times of the year when dust, dirt, dead insects, and excrement collect plentifully on the roof. The only impurities are those brought by the water itself if the cistern is made watertight with a close-fitting cover.

Arrangement should be made so that the first few minutes' fall of rain which has washed the atmosphere and washed the roof from which the water is collected shall be directed to waste, and only the comparatively pure water falling after this time shall be allowed to enter the cistern. The cistern should not be above ground, for in the summer months the water will warm up to such a degree as to promote the growth of bacteria in the water rather than retard it as is done with a cool storage. The use of a sand filter, such as is described in another article, is to be highly recommended where cistern water is used for household purposes.

It is not only the farmer who should be interested in pure water supplies for the farm, it is every man who is dependent upon him for milk and produce. Great typhoid epidemics are frequently traced to unsanitary farm conditions, which, when once pointed out to intelligent farmers, are quickly remedied, for there is no class of workers in this country more anxious to better living conditions socially, morally, and economically than the farmers of to-day.

CHAPTER IX

Running Water for Fifteen Dollars

MANY country homes on the farms of otherwise up-to-date progressive farmers still lack the great blessing of running water in the kitchen. Often this lack is due to a belief that the installation of such a supply would necessitate the outlay of a large sum of money. Nothing is more untrue. A really good and efficient running water system may be installed in the kitchen for less than fifteen dollars on almost any farm where there is a water supply, if the farmer is willing to do the work himself.

By reference to the drawing it will be seen that the supply is assumed to be from a well or cistern near the house. A pipe leads in through the cellar wall and below the frost line, then up through the first floor where a small cistern or tank pump is located. From this force pump the pipe line leads through a check valve up by the sink to an elevated tank which may be on the second floor, in the attic, or even on brackets just above the tank. A barrel makes a very satisfactory tank, or, where more water is needed,

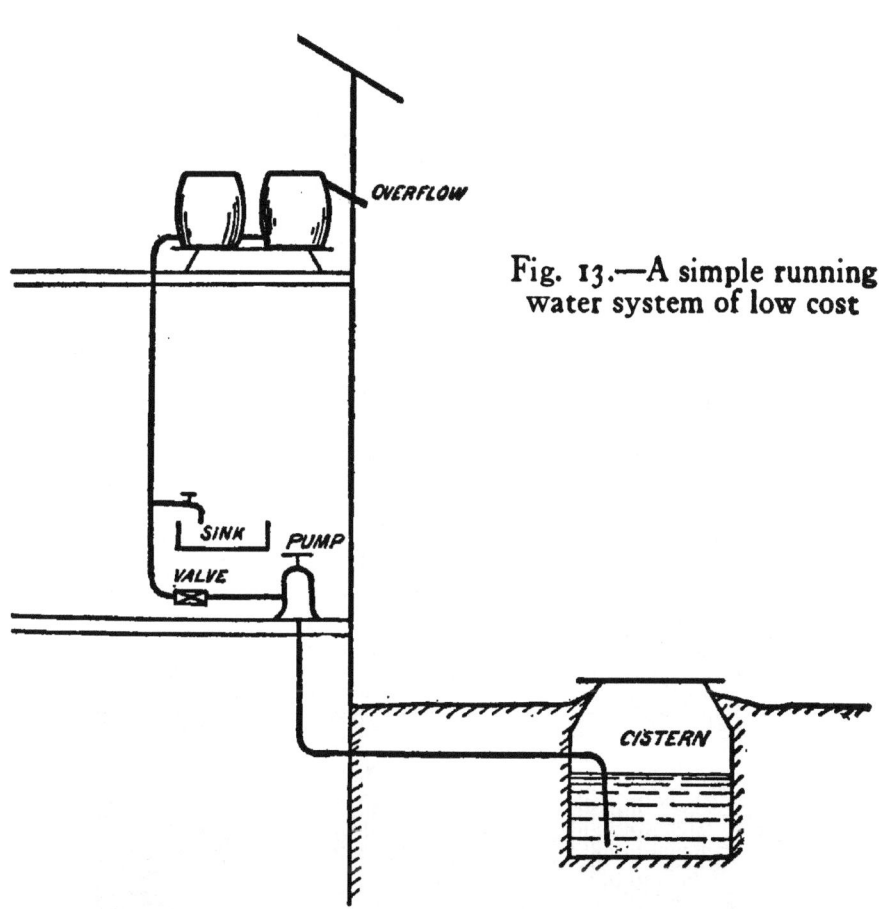

Fig. 13.—A simple running water system of low cost

several barrels standing side by side and connected at the bottom with short lengths of pipe. On a farm in northern New Jersey a farmer used six barrels elevated only about six inches above the sink and placed on a shelf in the pantry, the supply pipe to the sink faucet going through the pantry wall to the sink on the kitchen side.

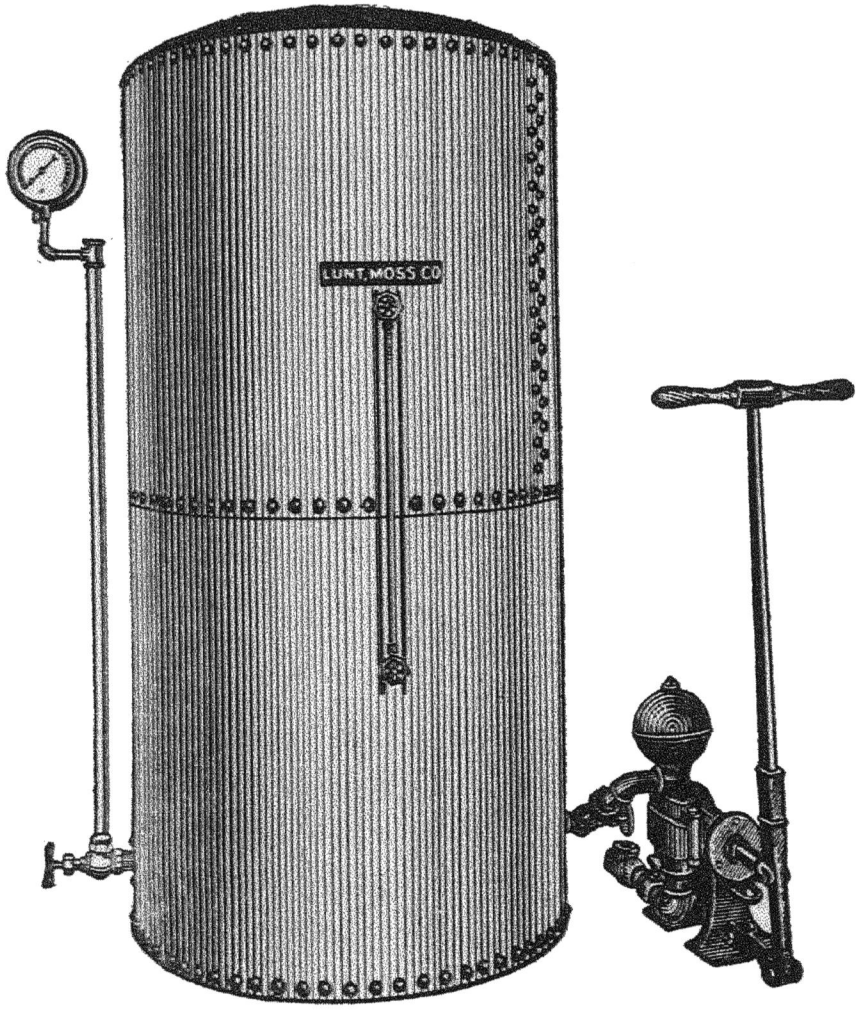

Fig. 14.—A simple pneumatic equipment

The pump recommended is the common, low-down, single or double acting force pump.

Fig. 12.—A neat and desirable spring housing

It will cost from $6 to $8. The check valve just beyond the pump will cost about 65 cents. It is a particularly valuable accessory, for it allows the water to flow from the pump to the tank but will not permit it to flow out in any way except through the sink faucet. The result is that the pressure of the water in the tank is not continually on the pump piston. If desired, a second pipe to another sink or to a handy faucet over the stove may be led off from any point between the check valve and the tank.

A good tank is formed by securing a barrel such as oil is shipped in, burning out the oil, and thoroughly cleaning. In the barrel place a small board as a float and run a string over the edge of the barrel to some point easily seen. Then hang a weight on the end of this string and tack up a paper or board marked as an indicator so that by the position of the weight you can tell how near empty the barrels are and whether or not a new supply should be pumped. One-inch pipe is large enough for all uses and one-half-inch pipe will give satisfaction for the stretch between the pump and the tank. It is best to use galvanized iron pipe, although black pipe is cheaper and does well if kept painted.

If a little more money may be invested this system may be readily enlarged to give running hot water, as shown in the next figure. A pipe

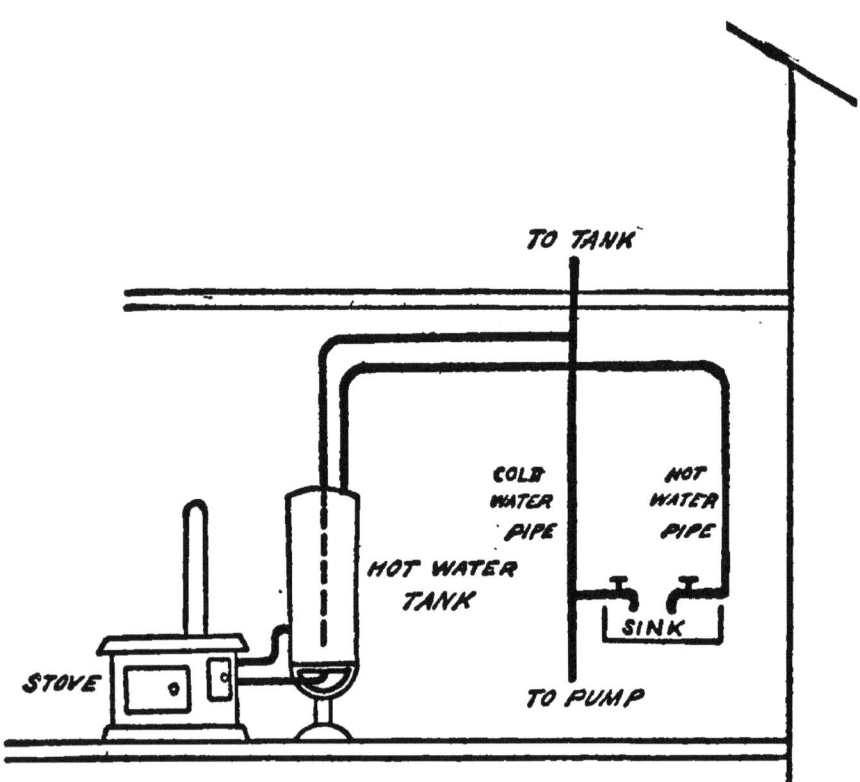

Fig. 15.—Illustrating the simplicity of the hot-water system in connection with the cold-water arrangement of Fig. 13

is led from the cold-water pipe to the bottom of a thirty-gallon galvanized iron tank costing about $5. From the top of this tank the hot-water pipe goes to the sink. In the stove is placed a coil of pipe, called a waterfront, and costing about $3.50. It is piped to the tank, one end going to the bottom and the other connecting about eighteen inches above. The tank will be drilled for connections, and the connections furnished for the price given above. The water in the waterfront becomes heated, rises through the coil, passes into the boiler, and rises to the top where it may be led off through the hot-water pipe. Meanwhile, cold water from the supply has passed down through the lower connection into the waterfront. This circulation continues over and over so that finally the tankful of water becomes hot.

It is readily seen that branch pipes may be led from these hot and cold water pipes to any part of the house below the storage tank. If the tank is in the attic, it is possible to have a bathroom on the second floor and have set tubs in the cellar for washing at only the extra cost of the tubs themselves and enough pipe to lead the water to them. Little by little, in this way, starting from the simple system shown in the first drawing, a splendid water supply arrangement may be built up, adding greatly to the ease and convenience of doing the daily tasks as well as adding greatly to the cash value of the house.

CHAPTER X

A Sand Filter for Rain or Brook Water

The use of screens, whether of wire or cloth for straining the water supply obtained from brooks, springs, and falling rain or snow is extremely unsatisfactory because of the ease and frequency with which they become clogged. Moreover, silt and fine particles are not removed from the water. The sand filter not only strains out the finest particles of suspended matter but also it has been found by careful investigations the water is purified chemically and bacteriologically. To a certain extent the filter allows thorough contact of the water particles with the air as the former trickle over the surface of the sand grains.

Usually the water is led to the top of the filter and allowed to seep down through the layers of sand and gravel to the lower part of the container, from which a pipe leads to a storage basin or reservoir. The house supply is pumped from the latter. If rain water is the source of supply, it is usual when no filter is used to allow the first few minutes' fall to run to waste in order that the impurities washed from the atmosphere and from the collecting roof area may not enter the storage basin. If a sand filter be used, this need not be done, although it is very advisable, for there is no advantage in having the filter do more service than is necessary. An automatic device may be used with safety, however, to divert the first fall.

One acceptable form of filter is shown in the diagram. There is a receiving barrel, a filter barrel, and a storage receptacle. The receiving barrel is in such a position as to receive the water directly from the roof and pass it out through a smaller pipe to the top of the filter barrel. In this way no more water is fed to the filter than can percolate through the sand even if the flow from the roof is very plentiful. If brook water is used, the receiving reservoir can be omitted and a pipe laid from the brook to the filter, or the filter may be made in a watertight container which is buried in the brook to such a level that the surface of the brook water is always slightly above the top of the container. In this way water is being freshly supplied to the filter at all times. A pipe from the bottom

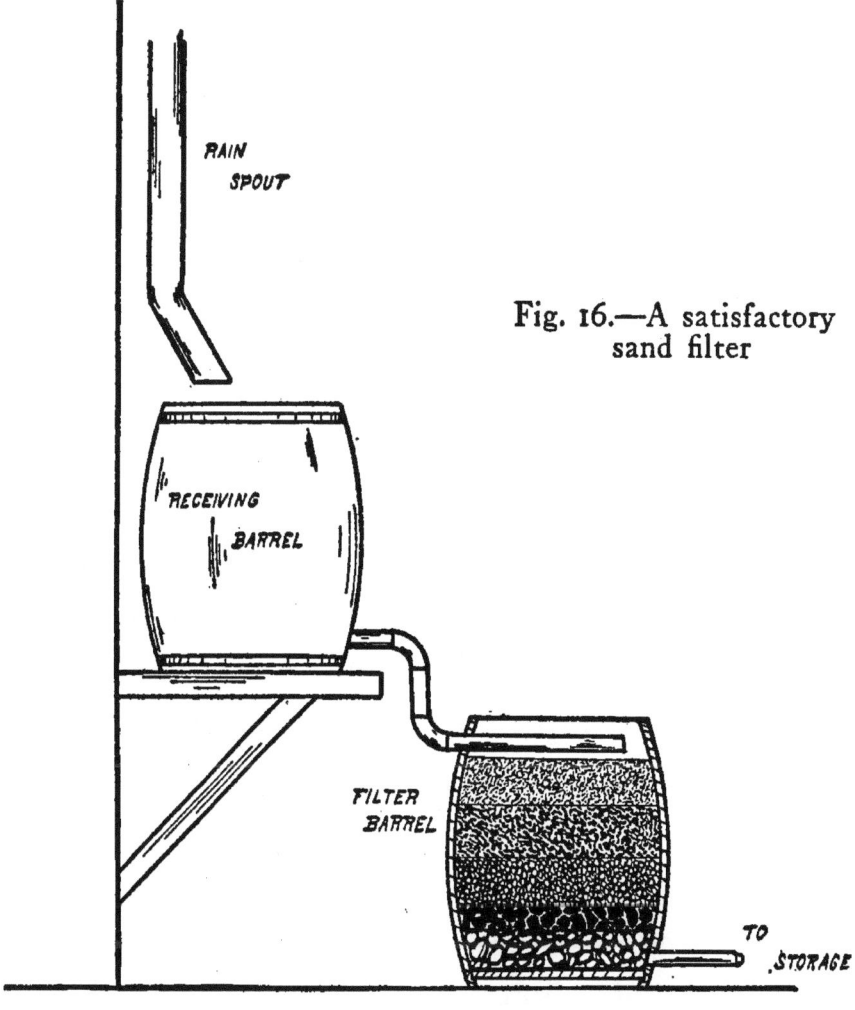

Fig. 16.—A satisfactory sand filter

of the filter leads to the main storage basin. As many receiving barrels as desired may be joined together, and more than one filter barrel may be used if it is desired to filter the water fast.

At the bottom of the filter barrel put a four-inch layer of coarse gravel and on top of that a layer of lump charcoal. Follow this with three layers of sand each ten inches thick, the first layer coarse, the next finer, and the top layer quite fine. Level each layer off well before putting in the next. Both sand and gravel should be clean and free from dirt and loam. It may be necessary to wash them before using. The flow of water to the top of the sand should be arranged so as not to disturb the layer. About three times a year (not oftener) the top four or five inches of sand should be scraped off and replaced by a similar amount of clean, fresh sand. Even this top layer must not be of the extremely fine sand sometimes found, although it is desirable to grade the layers.

CHAPTER XI

SOFTENING HARD WATER

THE carbonates and sulphates of lime and magnesia when present in water produce the effect known as hardness. This term applies merely to the difficulty with which a lather is obtained by using the water with soap. It is really of interest therefore only in so far as it affects the use of the water for washing purposes. So much of the water obtained in the country is hard water that a method of softening it should be of interest to every one who lives in rural districts. The "hardness" in the case of well water is usually due to the limy or chalky character of the soil through which the water flows, the water dissolving some of the lime content of soil. The hardness is noticed because the salts present in the water decompose the soap and form a sort of curds instead of a real lather. If more and more soap is used the whole of the troublesome material is used up and then a true lather may be formed without difficulty. That is, one way of softening the water consists in the plentiful use of soap.

If the hardness is caused altogether by carbonates spoken of above, it may usually be entirely removed by boiling for a short time. In this case the acid is driven off with the steam and a precipitate is left in the water which may be filtered off by pouring the water through a fine cloth or very fine screen. If, however, the hardness is caused by the sulphates, it is what is known as "permanent" hardness and cannot be removed by mere boiling as "temporary" hardness can. The most common and effective way of removing permanent hardness is by the addition of carbonate of soda, usually called washing soda. This causes the formation of carbonates instead of sulphates, and the carbonates may then be removed by boiling and filtering. Frequently borax or ammonia is used in place of washing soda.

There is another rather interesting way of removing the temporary hardness. It is by the addition of lime water (which is quicklime dissolved in water) or by the addition of a little lime. That is the queer thing. By adding a little lime to the water you get rid of the lime already there. The fact is that the lime in the water is held there because of the excess of carbonic acid. When more lime is added, this acid is neutralized and its effect is lost so that all of the lime is then precipitated and may be strained or filtered off by passing the water through a cloth.

CHAPTER XII

The Hydraulic Ram and the Ram-pump

To the average person the hydraulic ram is a mysterious thing. Working day and night for years without attention and without rest, it is the farmer's most dependable friend for pumping water. The efficiency of the ram when used for lifting water only four or five times as high as the fall is as great as that of the best pumps, and is much better than that of most pumping apparatus. For other ranges where the lift is from a small value up to twenty-five times the fall, the following table gives the efficiency of a ram:

TABLE A

Lift divided by fall	2	3	4	5
Per cent. efficiency	90%	85%	80%	75%

Lift divided by fall	10	15	20	25
Per cent. efficiency	57%	42%	30%	23%

The efficiency of a ram falls off so greatly as the delivery height increases that rams are seldom used where the lift is more than twenty-five times the fall. For what are known as "common rams" the general rule for calculation is that one sixth of the water supplied to the ram will be lifted to a height ten times as great as the fall. Exact calculation may be made for any ram by using the formula:

$$q = \frac{Q \times H \times e}{h}$$

where q equals the quantity of water raised, in gallons, Q is the quantity supplied to the ram, in gallons; h is the lift from ram to storage tank, in feet; H is the fall from supply down to ram, in feet; and e is the efficiency of the ram taken from Table A above, where h divided by H is the lift divided by fall.

For example, there is a fall of ten feet, and ram can be supplied with twenty-five gallons of water per minute. The storage tank is in the attic forty feet above the ram. How much water per minute will be supplied to tank? From Table A, the ratio of forty feet lift to ten feet fall will permit an efficiency of 80 per cent. Then, using the figures given and substituting them in the formula:

$$q = \frac{25 \times 10}{40} \times 80\% = 5 \text{ gallons per minute.}$$

It is apparent that if twenty-five gallons of water are delivered to the ram and only five gallons reach the tank, there must be a great waste of water. The water is wasted but the energy of its fall is utilized in lifting the remaining quantity to the greater height.

TABLE B

SUPPLY REQUIRED TO DELIVER ONE GALLON PER MINUTE

Ratio of lift to fall	2	3		5
Gallons per minute required to operate ram.	2.22	3.47	5.00	6.67
Ratio of lift to fall	10	15	20	25
Gallons per minute required to operate ram.	17.54	35.91	66.67	108.70

A diagrammatic form of ram is shown in the drawing. There are five main parts, the drive pipe A, the waste valve B, the delivery pipe D, the air chamber C, and the admission valve E. The water flows down A and out of the waste valve B when the ram is first started. When sufficient velocity has been gained by the water, it closes valve B suddenly. This confines the water in the casing and, as the movement of such a large bulk of water cannot be stopped

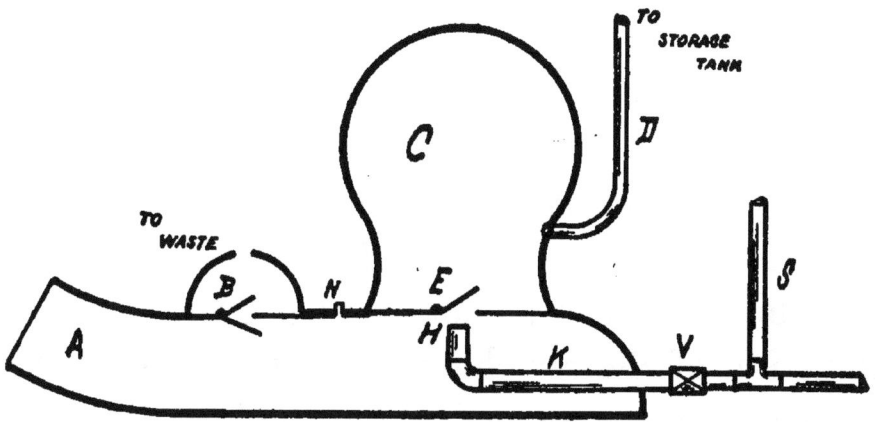

Fig. 17.—Diagram showing parts of ram and ram-pump

instantaneously, the valve E is dealt a hammer blow which opens it and allows a small amount of water to flow into the air chamber. The valve E then falls shut again and, too, as the water has slowed down, the waste valve B again opens, the water flows out, gains velocity, shuts the valve B again, opens valve E, and more water is forced into the air chamber. This action continues indefinitely as long as water is supplied to the ram.

The presence of air in the chamber C is necessary, for it compresses when the sudden blow is struck on the valve E, and this allows that valve to open. Of course the water will absorb a little of the air and after a time the air in the dome will be exhausted. This will cause the ram to stop and to prevent such stoppage there must be a way of admitting more air into the air chamber. This is done by boring a small hole at N. The water rushing into chamber C sucks in through the hole N just a tiny bit of air, but enough to prevent the exhaustion of the air chamber. On many of the higher priced rams a "sniffer" valve is located at some such point as N to serve the same purpose as the tiny hole here recommended.

The ram as described above will raise a portion of the water supplied to it to any desired height. If, however, it is desired to pump clear water from a brook or spring by means of undesirable water from some pond or stream, it may be done with safety by using a ram-pump. This resembles the ram shown except for the addition of the parts K, S, V, and H, as shown in the figure. As before, the water to operate the ram comes through the drive pipe, but the water to be pumped enters through the small pipe K and passes through the valve E when the latter is opened. A check valve at V prevents the clear water being forced back up the pipe K while a stand pipe at S keeps sufficient water pressure on the pipe at H to fill the right-hand end of the casing at all times and even allow a little to leak through the waste valve B. Thus, none of the impure water gets near enough to the valve E to be in any danger of being forced into storage. The ram-pump is best used where the supply of pure water is decidedly limited in quantity.

Rams and ram-pumps are usually placed at the bottom of pits dug into the ground, the head being increased in that way while the waste water flowing from the waste valve is easily drained from the pit through open joint tiles or through a drain pipe laid from the pit to a lower level.

Rams are commonly made in six sizes, from that requiring only one and one half gallons per minute to operate it up to one requiring twenty-five gallons per minute. The price ranges from $5 up to $25 for these sizes. Larger sizes are made, and often a whole battery of rams are installed where the supply of water is large. Ram-pumps are slightly more expensive. If possible, the ram to be purchased should be provided with an adjustable arrangement on the waste valve so that the latter will not stick if a higher head is used than was at first thought to be possible. If this is not done, care must be taken that the ram bought is workable on the highest head of water that **can** be used by you.

CHAPTER XIII

Disposal of House Sewage

ALLOWING sewage to flow directly into a stream or even into a cistern without first removing the harmful content is a serious mistake, and in many states there has been successful agitation for punishing such an act by fine or imprisonment. The most practical method yet devised for the disposal of house sewage without troublesome care and constant attendance is the septic tank method. It depends for its value upon the action of certain bacteria already present in the sewage. The conditions are made best for the growth and work of these bacteria, and they are permitted to liquefy and destroy the solid matter in the sewage. After their action the liquefied remainder is disposed of readily on any farm without giving cause for offence. Such a tank will not freeze in the coldest climate if buried a foot in the ground and used daily. No disinfectants are used.

It will not contaminate a nearby well or spring if the tank is made waterproof by plastering the walls with cement mortar.

The bacteria utilized are of two kinds: Those known as anaerobic thrive and grow in darkness away from fresh air. They are permitted to get in their work on the sewage as it first comes from the house, being led into a tightly covered, watertight, non-ventilated underground tank, and permitted to remain there undisturbed for

Fig. 18.—The septic tank

twenty-four hours. At the expiration of this time it is almost entirely liquid, and may be led over a filter bed of gravel or a well-drained trench filled with stone. Here the other variety of bacteria, called aerobic, assisted by the oxygen of the air, transform the murky liquid into a perfectly harmless substance which may be permitted to flow over the surface of the land, or may be discharged into a stream without any danger whatever of contamination.

One of the best forms for the septic tank to take is that shown in the illustration. It consists mainly of a concrete box three feet wide, eight feet long, and three feet deep. Three feet from one end is placed a partition which is perforated at a number of points in order that the liquefied sewage may pass through without agitation of the entire contents. The inlet pipe must be below the level of the sewage, as it stands in the tank and the perforations spoken of should be on about the same level. The outlet is somewhat higher than the inlet, but as the inlet pipe slopes from the house down to the tank, the outlet will be below the upper portion of the inlet pipe, and thus the tank will overflow properly. It requires some time for the tank to get to working in a thoroughly satisfactory manner, but after a little while a thick scum forms on the top and must not be disturbed or broken up. That is the main reason for introducing the inlet pipe below the surface of the liquid.

There is a hole left closed with a removable but tightly fitting cover in the top of the main chamber in order that the settlings at the bottom may be removed if found necessary after a few years' use. Under no other circumstances should the contents be disturbed. These settlings, if any are present, are mainly mineral matter, not from the sewage itself but from the paper or other foreign substances which enter the tank. The probabilities are that it will not need cleaning out for ten or fifteen years.

The tank should remain full up to a certain height at all times, this height being such that all sewage will remain in the tank about twenty-four hours or slightly longer. By placing an outlet leading to the filter bed at the right height, it may act as an overflow for the liquid, thus doing away with any necessity for watching and operating a valve. The outlet usually comes about twelve inches below the surface of the liquid. The inlet is usually about twelve inches above the bottom of the tank. As shown, the inlet should point downward inside of the tank as a further guard against undue disturbance of the contents.

The filter bed consists of another concrete box filled with stones and gravel in order that the liquefied sewage may trickle over it slowly, coming in contact with the oxygen of the air and allowing the aerobic bacteria to render the fluid harmless. From the bottom of this filter bed the purified sewage may be discharged to any convenient place. The usual way is to let it pass off through a four-inch tile drain fifty or sixty feet long set with open joints. The filter bed should be well exposed to air and light. The sewage when flowing from the bed should be clear, free from odour, and should not contain any poisonous or otherwise harmful matter.

The essential thing is to understand the simple theory of bacterial action which lies back of the septic tank process. If that is once firmly grasped, the details of tank building may be widely altered to meet particular needs. One very successful modification of this scheme, which has now been in use for several years, consists of simply the first tank spoken of above, and no partition in it, but simply an overflow arranged at the proper height to empty into a number of tiled drains laid out in the form of a network around the tank and about a foot beneath the ground. In this way the bacteria in the upper soil layers do the final work of purification.

The cost of such a tank as illustrated here

should not exceed $15 or $20 dollars, being near to the former figure if the cost of labour is not included, and near to the other figure if labour must be paid for. This estimate includes the purchase of cement, sand, and gravel, the lumber for the forms, the tile for the drains, and a hundred feet of vitrified sewer tile for the inlet pipe.

PART III

FARM POWER

Kerosene, Gasoline, and Coal as Fuels.

The Oil Tractor on the Small Farm.

The Ignition System and Ignition Control of the Gasoline Engine.

Determining the Horsepower of an Engine.

Utilizing Small Streams for Power.

The Storage Battery for the Farm.

CHAPTER XIV

Kerosene, Gasoline, and Coal as Fuels

The determination of the relative values of gasoline, kerosene, and coal for small engines has occupied much thought during the last few years because of the constantly increasing price of gasoline and the much cheaper cost of kerosene in many parts of the country. The problem is somewhat complicated, however, because it is necessary to take into account the relative costs of attendance and repair of the engines.

Kerosene is a heavier distillate than gasoline, both being obtained from petroleum. Theoretically, kerosene has a higher heat value than gasoline in the proportion of 11 to 9, but it is difficult to obtain the full heat value of kerosene in a gasoline engine. A much higher temperature is required to vaporize it than gasoline, and more evaporating surface is required. Then, within the engine, combustion is not apt to be complete, so that a deposit of carbon is left on the cylinder walls, piston, and spark plugs, thus requiring frequent cleaning. Any gasoline engine will run on kerosene if started and warmed up on gasoline, and the cost of fuel is less than when gasoline is used provided that three gallons of the former cost no more than two gallons of the latter. For example, using these fuels on small tractor engines the cost of gasoline at 30 cents per gallon was 70 cents per acre, and using kerosene at 15 cents the cost for fuel was 50 cents per acre. As this shows you, the amount of kerosene used ordinarily is over one and one fourth gallons to one gallon of gasoline. In an engine built to consume kerosene, the kerosene effects a greater saving because of the more complete combustion. Under those circumstances, for work requiring the same power for the same length of time, trials with small engines have shown an actually smaller consumption of kerosene than of gasoline, the latter being used in a gasoline engine.

In general, the amount of fuel consumed per horsepower per day of ten hours using gasoline is about one gallon, sometimes more but seldom less. The amount of coal consumed in the average farm steam engine per horsepower per day varies from sixty to eighty pounds with ordinary firing, although an expert fireman could probably cut that down to forty pounds. With this as a basis, you can figure the cost of fuel in your locality. Say gasoline is 30 cents a gallon and coal is $5 a ton, the cost of gasoline for a ten-horsepower engine per day would be $3, while the cost of a steam engine giving the same power would be (using sixty pounds per horsepower day) $1.50. If kerosene at 15 cents a gallon is used in the gasoline engine, the cost would be about $1.88. It must be remembered, however, that coal will have to be used for an hour or so getting up steam, and the coal on the grate at the end of the day, as well as the heat in the boiler at that time, is all wasted. The cost of the steam engine, therefore, for fuel alone will probably be nearer to $2 than $1.50. Again the steam boiler will mean added cost because of the constant attendance necessary in keeping up the fire and keeping the water level in the boiler from getting too high or too low. It will require attendance in getting up steam before time to use it, and it will need nearly as much attention between spells of using the engine, if the fire is kept up, as when the engine is used.

There are many other points in favour of the oil engine, such as less chance of explosion, less complication, and less knowledge necessary to operate it. The interest on the first cost of the plant will be less, and the depreciation should be slightly less. Yet the steam engine has a number of points in its favour. It can be "overloaded." That is, by increasing the steam pressure in the boiler, the ten-horsepower engine can be made to give twenty or twenty-five horsepower for a time, with increased coal consumption, of course. The exhaust steam may be used about the farm to heat water for washing purposes, or, if properly arranged, the exhaust from a stationary engine may be used to heat outbuildings.

There is an economic importance in the use of kerosene as a fuel in place of gasoline, for it will tend to lower the price of gasoline by lessening the demand. While, of course, by the same argument it will tend to increase the price of kerosene, there is such an oversupply of the latter due to its production in great quantities as a by-product in the refining of gasoline and other oils, that this tendency will not be very much felt.

All of these things being taken into consideration, there is no doubt as to the greater value of the kerosene engine. This is being demonstrated by the constantly increasing use of it, and a similar constantly decreasing use of steam engines in proportion to the total power used.

CHAPTER XV

The Oil Tractor on the Small Farm

While it is certain that the horse can never be entirely dispensed with on the small farm, the light-weight oil tractor of from six to thirty-five horsepower capacity is destined to relieve him of much of the hard work which he does but slowly and which wears him out in the doing. In many places where a horse is valuable the tractor cannot be used, but the decreased cost per acre of farming with the small tractor over that incurred when using horses; the fact that the tractor enables the farmer to do without help at just the time when help is scarce; the fact that when idle the tractor costs nothing to keep; that it requires no rest even on hot days, but, in emergencies, can be used all day and, with lights, work continued after dark; that, being small, it is economical in doing many things besides plowing—can, in fact, do all that a portable engine can do and, besides, propel itself where-

Fig. 19.—A tractor in the lumber country

Fig. 21.—A caterpillar tractor working in ground after plowing

ever it is wanted; all these advantages mean more money to the small farmer using such power. He is facing a new era in agriculture. His land, in many cases, has been so abused in the past by the continuous growth of crops without fertilization that it will require hereafter as much plant food put into the soil as is taken out in the crops. This means an added amount of work each year which must be done at certain times. Labour is no longer cheap, and satisfactory farm help is hard to get at any price.

The average work day of a farm horse the year round is only from three to four hours. Yet he must be fed the whole year at a cost averaging, perhaps, $100 for the twelve months. It may not be in cash, but in food that would sell for such an amount if there were no horse. His field of work is limited. Most of the small machinery which runs by belt power is not satisfactorily operated by either a horse sweep or a treadmill, while feed cutters, silo fillers, threshers, and similar machines are too heavy for the horse to handle. His speed on the road under load is very limited, as is his pulling power. The time taken in his care, the repairs to harnesses, the hitching and unhitching several times daily, allowing rest when work is waiting, are all features which raise the operating cost of a horse to a high amount per hour of work he does. On the other hand, the small tractor can be worked continuously every day to plow, harrow, drill, disc, harvest, haul, thresh, run small machines or the largest apparatus, and when it is not needed the power is immediately shut off and costs of operation cease.

The light-weight oil tractor can work in much softer soil than the larger sizes or even steam machines of the same rating, and on the road it can cross bridges and culverts which would need reinforcement before the heavy steam vehicles or the high-powered gasoline tractors could be taken on with safety. The following tables give many useful facts concerning the smaller size machines, and opportunity is thereby given to compare the steam tractor with the small gasoline types. Unfortunately, the rated horsepower in the case of steam tractors is not actual horsepower which can be developed, but is approximately half that maximum value. An oil tractor, then, should be compared with a steam machine of half the rating.

TABLE 1

TRACTORS OPERATING ON GASOLINE OR KEROSENE

Brake horse-power	Weight in lbs.	Drawbar pull in lbs.	Economical load in lbs.	Horse equivalent	Cost in dollars
6	3,086	900	4,000	3	$ 600
8	3,115	1,416	6,000	5	725
12	3,275	2,124	10,000	7	800
18	5,025	3,200	14,000	10	1,000
25	7,500	4,000	18,000	13	1,500
35	11,500	6,000	26,000	20	2,000

TABLE 2

STEAM TRACTORS

Horse-power ratings	Weight in lbs.	Drawbar pull in lbs.	Economical load in lbs.	Horse equivalent	Diameter of turning circle in ft.
15	14,000	4,500	20,000	15	35
20	20,000	7,500	26,000	25	40
25	21,000	9,000	30,000	30	40
30	28,000	10,500	40,000	35	45

The above tables are representative and include a number of different makes of tractors, all of which are guaranteed for one year against defects of manufacture.

In determining the size of tractor needed for any particular work, the following tables will be of interest. In reading them, it should be observed that drawbar pull corresponds to the pull on the traces by horses or the pull on the pole by oxen. It is the pull which the tractor will exert through the medium of the drawbar, the bar which couples the tractor to the train it draws after it.

TABLE 3

DRAWBAR PULL REQUIRED FOR A LOAD OF ONE TON IN WAGON

	Good road *lbs.*	Gravel *lbs.*	Sand *lbs.*
On the level	125	250	625
Rise of 1 foot in 100 feet	145	270	645
" " 2 " " " "	165	290	665
" " 3 " " " "	185	310	685
" " 4 " " " "	205	330	705
" " 5 " " " "	225	350	725
" " 6 " " " "	245	370	745

TABLE 4

*DRAWBAR PULL REQUIRED FOR ONE PLOW BOTTOM IN VARIOUS SOILS

	12-inch bottom		
	6-inch *lbs.*	7-inch *lbs.*	8-inch *lbs.*
Sandy	216	252	288
Clover sod	504	588	672
Clay	576	672	768
Virgin sod	1,080	1,260	1,440
Prairie sod	1,080	1,260	1,440
Gumbo	1,440	1,680	1,920

*Each plow will turn about 2½ acres a day at 2½ miles per hour.

Compliments of the Holt Manufacturing Co.
Fig. 22—A caterpillar tractor working in swamp land

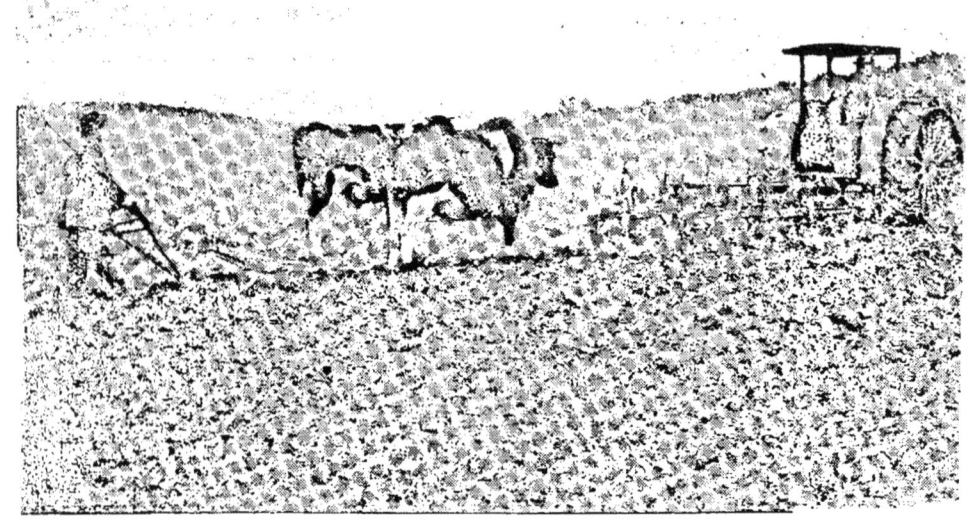

Fig. 23.—The past and the present

Fig 24.—A severe test for any machine

TABLE 4—Continued

	14-inch bottom		
	6-inch lbs.	7-inch lbs.	8-inch lbs.
Sandy	252	294	336
Clover sod	588	686	784
Clay	672	784	896
Virgin sod	1,260	1,470	1,680
Prairie sod	1,260	1,470	1,680
Gumbo	1,680	1,960	2,240

	16-inch bottom		
	6-inch lbs.	7-inch lbs.	8-inch lbs.
Sandy	288	336	384
Clover sod	672	784	896
Clay	768	896	1,024
Virgin sod	1,440	1,680	1,920
Prairie sod	1,440	1,680	1,920
Gumbo	1,920	2,240	2,560

In connection with Table 4, it should be noted that in going up a grade each rise of one foot in one hundred feet adds 1 per cent. of the weight of plows and tractor to the pull required, and in going down such a grade 1 per cent. is taken off the pull required. A plow gang weighs from 600 to 700 pounds for each bottom, so, for example, a five-plow gang would weigh between 3,000 and 3,500 pounds.

The cost of operation should include not only the cost of fuel, oil, labour, and repairs, but should also include interest on the investment and depreciation in the value of the machine. The latter figure may be made to allow for repairs also. Interest on the money invested averages 6 per cent. the country over. Depreciation, including repairs, should be charged at 10 per cent. to be on the safe side. As to the consumption of fuel, most of the tractors require from one and one half to two gallons of fuel per acre of land plowed, but the time taken to plow an acre depends, of course, on the speed of the tractor and the number of bottoms pulled. The economical speed at which a tractor should run in the fields is from two to two and one half miles per hour. Slower than this leaves a poor job, while a faster speed does not permit the machine to exert its most economical pull. In drawing a load on the road, from three and one half to five miles per hour is a good rate of travel. For all operations requiring the use of the tractor engine as a portable engine, an allowance of about one gallon of fuel per horsepower exerted for a nine-hour run will be ample.

Average cost of operation, then, including

Fig. 25.—The one man outfit plowing

everything, and allowing a fair sum to the farmer for his time in driving and caring for the machine, should not be over 75 cents to $1 per acre of land plowed. The cost per working day for a twenty-five horsepower one-man oil tractor costing, say, $1,500, working ten hours per day for 200 days in the year, would be about as follows:

TABLE 5
AVERAGE COSTS OF OPERATION (APPROXIMATE)

Interest at 6% on $1,500	$ 90.00	
Depreciation at 10%	150.00	
Driver's time at $3 per day.	600.00	($5 in many sections)
Kerosene at 10c. per gallon	540.00	(average wholesale
Oil, grease, etc.	100.00	price)
Total for 200 days.	$1,480.00	
Cost per day	$ 7.40	

This machine at a cost of $7.40 per working day will do the work of seven two-horse teams and drivers, assuming that the horses and drivers could work ten hours per day for 200 days in the year. If hired, seven two-horse teams with drivers would cost at least $20 per day.

If the men were hired and the horses owned, the cost would be somewhat less than if the whole outfit were hired, as shown by Table 6, which follows:

TABLE 6
COST OF HORSES AND HELP

Assume first cost of horses and harnesses	$3,000.00
Interest at 6%	180.00
Depreciation at 10%	300.00
Feeding and care	1,400.00
Total for 14 horses	$1,880.00
Cost per day	9.40 (Assuming 200 working days)
Wages, 3 men, at $2	6.00
Cost per day, 7 teams and drivers	$ 15.40

Undoubtedly the figures given in these tables do not apply to every case. They are, however, of value as pointing out what things to take into consideration, and give some definite idea of prices that do obtain in some sections.

CHAPTER XVI

The Ignition System and Ignition Control of the Gasoline Engine

As SPRING approaches, thousands of gasoline and kerosene engines will be brought into service all through the farming districts as stationary and portable engines, operating all kinds of farm machinery, and as automobile, tractor, and truck-propelling engines. Two thirds of the difficulties encountered in their operation will be due to defects in the ignition systems, or to lack of knowledge of the importance of proper ignition control. The ignition system is the vital part of the oil engine, and it must work properly and be controlled in the correct manner.

There are two divisions of ignition systems under which all designs may be properly classified: the make and break or low tension, and the jump spark or high tension. These names refer to the particular method by which the spark in the cylinder is made. With the former design there are two contact points in the cylinder, one of which is movable and may be turned away from the other suddenly by a spring trigger arrangement, after having been in contact for a very small interval of time. The two points are connected in circuit with a battery and a coil of wire wound about an iron core. When the points are separated the "momentum" of the current causes it to jump the gap created between the points, thus giving the required spark. The purpose of the coil used is to increase this tendency of the current to continue to flow even after the circuit has been broken. The coil itself consists merely of a few turns of insulated copper wire wound about a soft iron core. Such an arrangement as this has been used for many years in electric gas-lighting systems and is there known as a spark coil. It is commonly referred to in connection with gas engines as a make and break or a non-vibrating coil.

The make and break system, because of the difficulties of mechanical design, cannot be used on high-speed engines nor on very small sizes. It has many advantages and many disadvantages over the jump spark system. A much hotter spark can be obtained with the make and break because of a greater flow of current. There is not so much leakage of current and it is not so readily put out of service by dampness and dirt. On the other hand, good contacts are required all through the system, and particularly the contacts within the cylinder must be kept clean. This is difficult, for there is a continual deposit of soot and oil. Mechanically, the system is defective because of the numerous moving parts and wearing surfaces.

The jump spark design is that in which a spark plug is used in the engine cylinders. Here the spark points are stationary (but adjustable) with a fixed distance between them. They are in circuit with the secondary of an induction coil, commonly referred to as a jump spark coil or a vibrating coil. It consists of two windings. The primary has a few turns of comparatively large copper wire and is connected to the battery. The secondary has many thousands of turns of fine wire, the fine wire being used solely to allow the coils to be crowded close to the core and to save space and cost.

Owing to the larger number of turns in the secondary, the voltage or "pressure" of that circuit is higher than that of the battery circuit, and so it can force a flow of electricity across the gap. The current flowing in the secondary is less than that in the primary, and it cannot be measured easily and directly by convenient instruments. The current in the primary, however, may be measured by means of a pocket battery ammeter, and should not exceed one fourth or one half an ampere if the circuit is in proper condition.

The principal disadvantage is the high tension or voltage used, because of the difficulty with which proper insulation is obtained. The least dirt or moisture is fatal to the workings of the system. The vibrator in the primary circuit used to rapidly open and close the circuit is many times the source of much annoyance. On the whole, however, this system is most generally adopted for medium and small sized engines.

Magnetos are common in ignition systems, the low tension replacing the battery in the make and break systems and, occasionally, in the primary of the jump spark design. The high-tension magneto, when used, takes the place of the whole jump spark system if desired, the spark plug being connected directly to it.

In all of these systems the electrical action is practically instantaneous, but it is not always realized that, although combustion in the engine cylinder is extremely rapid, there is a definite period of time which occurs between the closing of the electrical circuit and the point of maximum pressure set up by the explosion of the gases. Such is the case, however, the exact time depending upon the proportions of air and oil vapour in the mixture, as shown by the following table of approximate combustion periods:

TABLE OF COMBUSTION PERIODS

Mixture proportions	Time of combustion in seconds
1 part gas to 4 parts air	0.04
1 part gas to 7 parts air	0.08
1 part gas to 9 parts air	0.12
1 part gas to 11 parts air	0.18
1 part gas to 12 parts air	0.23
1 part gas to 13 parts air	0.28
1 part gas to 14 parts air	0.31

Because of this slowness of combustion, the spark circuit must be closed a little while before the piston gets to the exact point where it is desired that explosion take place. Sometimes, for example, the spark circuit is closed before the piston reaches the end of its compression stroke. Yet, at the same time, the force of the explosion does not occur until after the maximum compression has taken place and the piston started back.

There are, particularly with automobile engines, many changes from time to time in the richness of the mixture, and so, of course, there must be changes in the point of ignition because there will not be the same intervals between closing the sparking circuit and the point of complete combustion. This variation in the mixture is due to changing the throttle, opening and closing it from time to time as the load varies. Then, too, with an increase in the speed of the engine the spark must be advanced because the circuit must be closed earlier in the stroke to allow the same period of time to elapse before the piston reaches the end of stroke, the piston travelling so much faster than before. On the other hand, if the engine is being started, the piston is travelling slowly and so the spark must be retarded. That is, the circuit must be closed at the time when the piston is at the end of the stroke, or after it has passed the end of stroke, usually the latter. In either case the maximum force of the explosion will occur after the piston has started back. Care should be taken that explosion shall not occur when the piston is exactly at the end of stroke, because that causes bad knocking owing to the full force of explosion being transmitted directly to the crank and crank-shaft bearings.

If explosion occurs before the piston reaches the end of stroke when the engine is starting, it may reverse the direction of motion of the crank and so injure the operator who is trying to turn it over the other way. If the explosion occurs too early, when the engine is running, there will be a loss of power because the force of the explosion will oppose the motion of the piston. Then, too, combustion is slower with the gas under less pressure, so that the engine will become overheated if running continually with a much retarded spark.

These facts underlie three rules of spark control which should be memorized and understood by every engine operator:

1. Always retard the spark before starting the engine.

2. Always advance the spark as the engine picks up speed.

3. Always retard the spark when the engine slows down under a heavy load.

In every case when the engine is running, the object of spark control is to get an explosion at the moment when the crank has passed the dead centre and the piston has started back on the return stroke. This will give the maximum power and the most economical operation. An explosion at any other time in the stroke wastes fuel and injures the engine—from undue strain if before the piston reaches the end of stroke, and from overheating if after.

CHAPTER XVII

Determining the Horsepower of an Engine

THERE are two values which are known as the horsepowers of any engine. One is called "indicated horsepower," and usually written I. H. P., while the other is the "brake horsepower," written B. H. P. The indicated horsepower of a steam engine is the mechanical work done in a certain time by the steam acting on the piston. Some of that work goes to run the engine itself, overcoming the friction of the bearings and the drag of the moving parts, so that only a portion of the force exerted can be delivered to the belt pulley. The work which can be done by this portion at the pulley, in a certain length of time, is the brake horsepower.

To understand fully the methods of measurement of these values, we must understand the terms work, power, and energy as they are used in engineering. Any force which is exerted through a distance is said to do work. For example, an iron weight which drops from a height of two feet does work, because the force of gravity equal to the weight of the iron is exerted through two feet. The amount of work done is the product of the force in pounds and the distance in feet. The unit used is the foot-pound. If, then, the iron above mentioned weighs ten pounds the work done when it falls two feet is the product of ten and two, that is, twenty foot-pounds. Work is the product of force and distance.

Energy is the ability to do work. It is the capacity for work that a body or substance has, and is measured in foot-pounds just the same as work. The weight mentioned, before it fell, had the ability to fall and do twenty foot-pounds of work. Thus we say that it possessed twenty foot-pounds of energy.

Power is the rate of doing work. If an engine can do 33,000 foot-pounds of work in one minute, it is a one-horsepower engine, that figure being the standard chosen to represent a horsepower. Power has to do with time. Any engine can do 33,000 foot-pounds of work, even a toy engine if you give it time enough.

The point to be noticed is that a one-horsepower engine must be able to do that much work

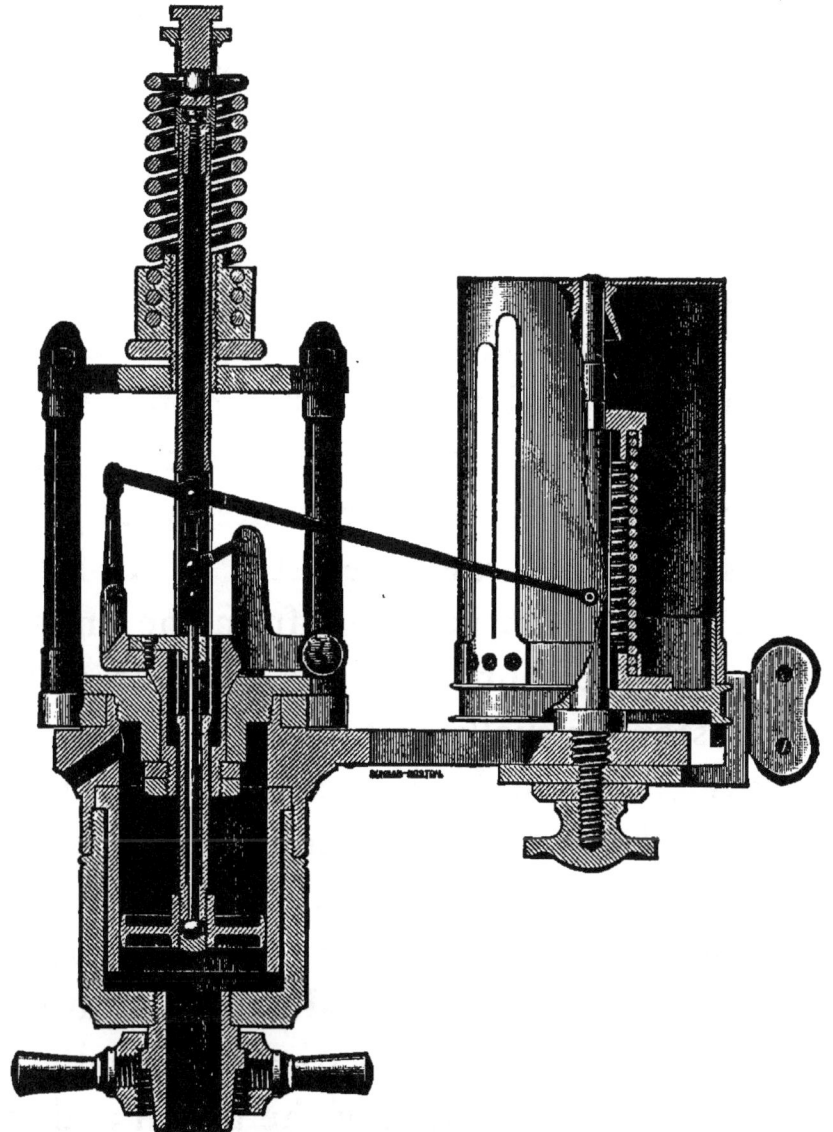

Fig. 26.—The engine indicator

in one minute. A two-horsepower engine must do that amount in half a minute, or, what is the same thing, it must do twice that amount in one minute.

To measure the indicated horsepower of a steam or oil engine, an instrument known as the indicator is used. The illustration shows the general appearance of the indicator used with steam engines, and the same general arrangement is found in all indicators. There is a cylinder to which steam is admitted from the engine cylinder. The steam forces the piston back against the resistance of a coiled spring which has been experimented with previously, so that the pressure exerted by the steam on the little piston is known from the amount the spring is compressed. As the area of the small piston is usually just one square inch, the pressure indicated by the compression of the spring is the pressure per square inch of the engine piston. So if we multiply this indicated pressure by the total area of the engine piston, the result obtained is the total steam pressure on the engine piston. This varies continually on account of the movement of the piston and the expansion of the steam.

There is also on the indicator a rotating drum which turns through a distance proportional to the stroke of the engine piston. A pencil is so arranged that it goes up and down with the indicator piston, and as the drum rotates beneath the pencil the latter draws a diagram with its length proportional to the engine stroke and its height proportional to the pressure on the engine piston. The accompanying figure

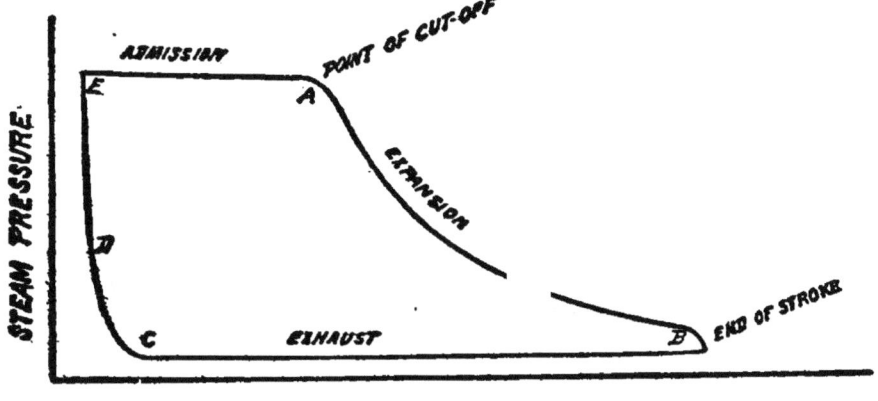

Fig. 27.—A typical indicator card. *A* is the point where the steam is cut off, *B* the point where the exhaust is opened, *C* is where compression begins, and *D* is where admission of live steam starts, the pressure rapidly running up to live steam pressure at *E*

shows the shape of such a diagram, and this is known as an indicator diagram. Mathematical calculations show that the area of such a diagram as this is proportional to the product of the average pressure on the piston during the stroke and the length of the stroke. In other words, the area of this diagram is proportional to the work done on the piston of the engine by the steam during one stroke, so that knowing the number of strokes per minute made by the engine piston we may easily find the work done per minute. This divided by 33,000 gives us the indicated horsepower (I. H. P.) of the engine, because 33,000 foot-pounds per minute equals one horsepower.

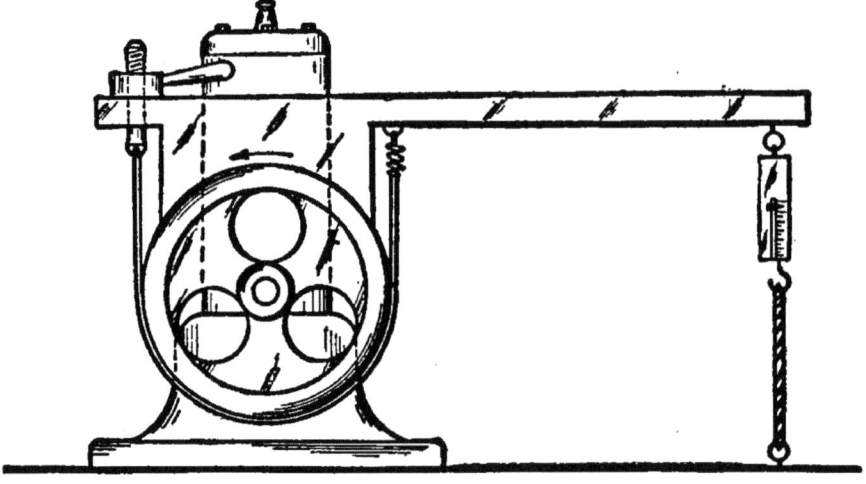

Fig. 28.—One form of prony brake

To measure the brake horsepower of any engine, an instrument known as the prony brake or, more technically, the absorption dynamometer, is used. This, as shown in the drawing, consists of a band which may be tightened around the engine pulley, creating great friction on the pulley and requiring constant force acting to overcome this friction. As this force is acting constantly on the rim of the pulley, in one revolution of the pulley the force acts through a distance equal to the circumference of the pulley. The circumference is three and one seventh times the diameter. The product of the length of the circumference and the force of friction acting will give the work done in one revolution. Then by counting the number of revolutions per minute and multiplying this number by the work done in one revolution and dividing by 33,000 we get the brake horsepower.

It is difficult to measure the force of friction directly, so that it is measured by suspending weights on the end of a long arm as shown in the figure. By the principle of the lever, the force acting at the circumference of the wheel is to the force exerted by the weights at the end of the arm as the length of the arm measured from the centre of the shaft is to the radius of the wheel or pulley. That is:

$$\frac{\text{Friction force}}{\text{Weights}} = \frac{\text{Length of arm}}{\text{Radius}}$$

and hence

$$\text{Friction force} = \frac{\text{Length of arm} \times \text{weights}}{\text{Radius of pulley}}$$

And the horsepower as stated above, being friction force times the circumference times the number of revolutions per minute (written R. P. M.) divided by 33,000 will give the following value for B. H. P. by substituting the value of the friction force found above:

$$\text{B.H.P.} = \frac{\text{Length of arm} \times \text{weights} \times 3\tfrac{1}{7} \times \text{pulley diam.} \times \text{R.P.M}}{\text{Radius of wheel} \times 33{,}000}$$

$$= \frac{\text{Length of arm} \times \text{weights} \times 3\tfrac{1}{7} \times 2 \times \text{R.P.M.}}{33{,}000}$$

because the diameter is twice the radius and we may divide them.

If now we take pains to have the length of the arm measured from the engine shaft to the weights, just $3\tfrac{1}{2}$ feet long, the formula becomes simplified and we obtain, by dividing

$$\text{B.H.P.} = \frac{\text{Weights} \times \text{R.P.M.}}{1{,}500}$$

The belt which creates the friction is usually made of heavy canvas held by springs at one end while a turnbuckle is used at the other end in order that the belt may be tightened at will and the force of friction increased. In long-continued tests it is frequently found necessary to throw water over the belt and pulley to keep them cool. In place of the weights a spring balance may be used, care being taken that the turning direction of the pulley is such as to pull against the balance. With small engines, up to ten or twelve horsepower, a balance reading to twenty-five pounds is large enough.

Gasoline and other oil engines for farm use are usually rated at their tested brake horsepower, but the power of steam engines, unfortunately, is not so accurately stated. The commercial power rating of steam engines is ordinarily only one half or one third of what they actually will do under test. The custom in making calculations is to assume that a steam engine of any specified rating will give the same power as a gasoline engine of twice the rating.

The rating of a steam boiler in boiler horsepower has nothing whatever to do with the unit of power we have been using. A boiler horsepower and an engine horsepower bear no definite relation to each other. A boiler horsepower is defined as equivalent to the evaporation of thirty-four and one half pounds of water per hour from water at 212 degrees Fahrenheit to steam at the same temperature and at the pressure of the atmosphere. Under ordinary conditions with farm engines, one boiler horsepower will furnish sufficient steam to operate an engine of about one half horsepower capacity. This, however, is only an approximation.

CHAPTER XVIII

Utilizing Small Streams for Power

The idea of water power is generally associated with a mental picture of an expensive installation which is beyond the purse of most farmers, yet it is no exaggeration to say that on many farms small streams could be harnessed to do the work required of an engine and with very little expense. The size of the stream and the amount of fall is of first interest, of course, in order that calculations may be made of the possible amount of power which can be generated. To understand how these calculations are made we must first find out just what is meant by "horsepower." This is discussed quite fully in Chapter XVII, and again in connection with Table VII. Power is defined as the rate of doing work, and the unit of power is taken as 33,000 foot-pounds of work done in one minute.

To find the maximum possible power which can be obtained from a falling body of water, then, it is only necessary to determine the weight of water which falls in one minute and the distance that this water falls. The latter distance, as a rule, may be easily measured. If the weight of the water is desired, the first step is to determine the quantity of water falling in one minute, usually in cubic feet. This quantity is obtained by multiplying the average depth in feet by the average width in feet and multiplying their product by the velocity of flow in feet per minute. It will assist the calculation greatly if a stretch of stream is taken which does not vary greatly in width. Say the stretch is 200 feet long. Measure the width of the stream at, say, ten places along this stretch. At each place take six measurements of the depth of the stream, spacing the measurements from shore to shore. Average the six measurements and multiply by the width at that place, getting the cross-section at each place. Average these cross-sections and multiply by the average velocity of the stream. The latter may be obtained by noting the time taken for a chip to float the stretch of 200 feet; or perhaps an easier method would be to note how far a chip will float over this course in one minute. The final result will be in cubic feet, and, as a cubic foot of water weighs 62.4 (Table III) pounds, multiplying by this figure will give the weight of water which falls in one minute at any part of the stream. Multiply this weight by the distance fallen through and divide by 33,000 to get the maximum theoretical horsepower. As stated in the chapter and table previously referred to, this maximum horsepower cannot be obtained from any commercial wheels, their efficiencies ranging from 50 to 85 per cent., as pointed out in Table II.

The term "miner's inch," which is sometimes met with in waterpower apparatus catalogues, is a California term and is really quite indefinite in its meaning, the exact amount meant depending upon the particular locality where the term is used. An average value for the miner's inch is one and one half cubic feet of water per minute. The number of miner's inches in any stream, then, may be approximately determined by dividing the stream flow in cubic feet per minute by one and one half.

If the water to be utilized is from a small waterfall, the calculations may be made in the stream above the fall. If necessary or desirable the stream might be led through an open wooden channel of known cross-section and known length. The width of the box will give the width of the stream. The depth of the stream can be measured by means of a rule held upright with the end resting on the bottom of the box. The velocity can be determined from the time taken for a chip to float the known length of the open box.

The following table gives the number of cubic feet which must fall per minute to give one horsepower under the various heads named:

WATER REQUIRED TO GIVE ONE HORSEPOWER

Head in feet	Cubic feet per minute required
5	105.6
10	52.8
15	35.2
20	26.4
25	21.1
30	17.6
35	15.1
40	13.2

Knowing the horsepower which it is possible to develop, the next thing is to choose the type of wheel. In general, water-wheels may be classified as gravity, impulse, or reaction wheels.

UTILIZING SMALL STREAMS FOR POWER 133

The gravity type are operated directly by the weight of the falling water exerted through its falling distance. Such are the breast and over-

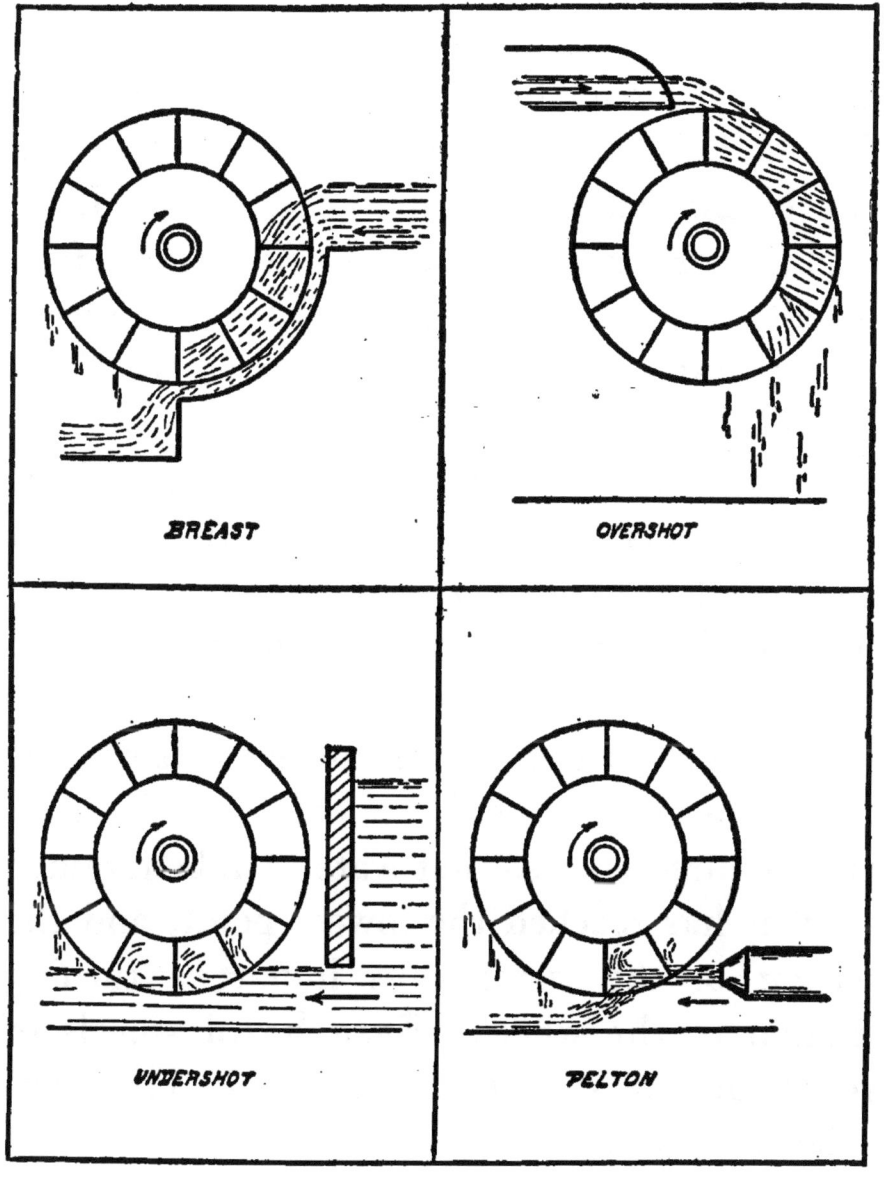

Fig. 29.—Types of water-wheels

shot wheels represented diagrammatically in Fig. 29. They are used solely in small plants, being inefficient under normal conditions. Under the best conditions the efficiency of the breast wheel ranges from 55 to 65 per cent., and that of the overshot from 65 to 75 per cent. If the fall in the stream is but a few feet the breast wheel is quite generally used. A slightly greater fall, say six to eight feet, usually results in the choice of an overshot. These gravity wheels are advocated for slight falls of from three to eight feet, or thereabouts, for small installations largely because of the fact that small turbines for slight falls are apt to be of low grade materials and poor design. The gravity wheels are much easier to make and install. In fact, overshot wheels are frequently constructed by the farmer himself. It may be any form of wheel with buckets or paddles on the circumference so that water will be retained until it has reached the lowest point, and the weight of the water thus impart a turning motion to the wheel. Even board wheels of rough design and construction will give considerable power.

Impulse wheels are those in which the total

energy of the wheel is obtained from the movement or velocity of the running water. The undershot wheel represented and the Pelton wheel are examples of this class. While the undershot wheel is perhaps the least efficient of all water-wheels, averaging from 25 to 45 per cent. under good conditions, the Pelton is the most efficient. Under favourable conditions the efficiency of the latter reaches 85 per cent., and in all intelligent installations it runs well over 75 per cent. A running stream having a slight fall furnishes opportunity for the common mill wheel of the undershot type. Where it is used the stream is narrowed to about the width of the wheel, thus giving the wheel the benefit of all the water in the stream running at a somewhat greater velocity than in the open stream. This type is rapidly disappearing altogether, and is not to be recommended if other types may be installed. Frequently in order to use another type as, for example, the breast wheel, a dam would have to be constructed to get a sufficient fall of water. There is a low breast wheel which is sort of a cross between the breast and the undershot. This is used where the fall is slight, say a foot or two. The water is delivered to some point of the wheel below the shaft, anywhere between one quarter and three quarters of the distance from lower point to shaft.

The Pelton wheel is increasing in use, and together with the turbine is universally installed in plants of any size. For all heads above eight or ten feet this wheel equals the turbine in efficiency. For heads less than twenty to twenty-five feet, however, the amount of water used by the Pelton makes the turbine somewhat more economical. Above twenty feet there is little choice from efficiency or cost of operation until high heads of from 100 to 2,000 feet are reached. With these there can be little choice between the two, the Pelton being greatly superior. The principle of operation makes a high head desirable with the Pelton wheel. The higher the head the less the amount of water required to develop a given power. Hence, the lower the cost of installation, for provision need be made to convey only a slight amount of water. The power of a Pelton wheel depends solely upon the head and the amount of water supplied to the wheel. The diameter of the wheel merely determines the speed at which it runs, and to some extent is dependent upon mechanical considerations. With great quantities of water flowing from the nozzle, the buckets against which the water strikes must be large enough for the full benefit of the issuing stream, and thus the wheel must be large enough to carry the buckets. Most of the so-called water motors are of the Pelton type. They range in price from $30 for the little six-inch motor, weighing fifty pounds, up to $275 for the twenty-inch size, weighing 860 pounds. Smaller motors down to about one eighth horsepower may be bought for as low as $10.

Turbines are of the reaction class of wheels, the reaction of the water as it leaves the vanes furnishing the "kick" which propels the wheel. In this type of wheel, in distinction from all others shown, the water acts around the entire circumference at once. The efficiency depends largely upon the design and the carefulness of installation. It may be anywhere from 55 per cent. up to 85 per cent. It is best adapted for low and moderate heads, especially where the head varies greatly from time to time. It operates at higher speeds than the other wheels, and will perform its work even if set below the level of the water in the tail-race. Low heads and large quantities of water cause the adoption of the turbine.

A preliminary survey and outline report by a competent engineer is advisable in every case where a waterpower plant of any great size is to be erected. Such advice is not expensive

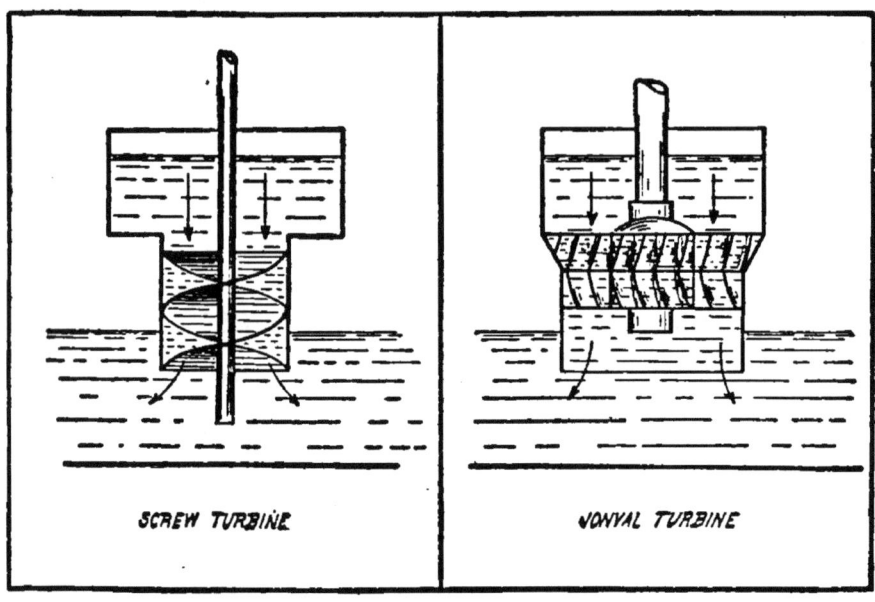

Fig. 30.—Diagrammatic representation of typical turbine wheels

and will many times set the farmer on the right track regarding details of his venture. For small installations, however, the farmer may rely on his own judgment and the help available from the manufacturer from whom he purchases the wheel. This chapter is written for the purpose of calling attention to the possibilities of the small stream running through the fields, and pointing out a method by which the farmer may calculate for himself the power available and the kind of wheel to install. Just what installation is best in each case and the exact cost depend upon local conditions. Before determining the size of wheel to use, the condition of the stream at all seasons of the year must be taken into account. The installation is for continuous use and average conditions must be figured on, for if the head of water is real variable a wheel too large for all but the highest heads will operate at a very low efficiency when the head is low. On the other hand, a wheel too small for any but the very low heads will have a low efficiency on the high heads. In almost every case the wheel is chosen to run at a certain fixed speed. This speed cannot be maintained under wide variations in head without affecting the efficiency of the plant. The usual solution is to arrange the plant so that the head will remain as nearly constant as possible and any surplus water go to waste. As has been stated, low heads are best developed by turbines and high heads by

Pelton wheels. These two types are practically the only ones which can be purchased for small installations. The other types, however, are quite readily built by an intelligent farmer.

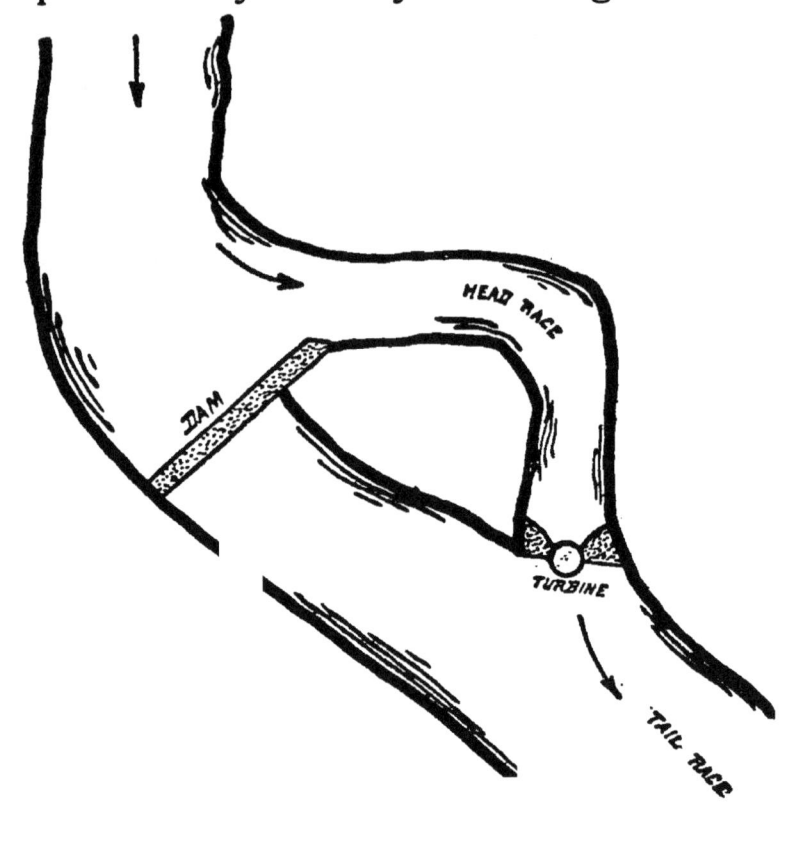

Fig. 31.—General location of dam and turbine wheel in most installations

For turbine installations the natural head at a fall is usually enlarged by building a dam across the stream at that point. Off to one side of the dam, as shown in Fig. 31, the raised water enters the head-race, goes through the turbine, and then goes out through the tail-race. The short length of pipe or open channel from the head-race to the wheel is called the penstock or flume. The portion of the watercourse in which the wheel is situated is called the wheel pit. The following table gives some figures about successful farm installations of waterpower in various parts of the country. All of these are turbine plants.

FARM WATERPOWER PLANTS

Head of water	Power developed	Length of dam	Cost of plant
6 feet	17 h.p.	36 feet	$1,000
11 "	8 "	350 "	1,000
15 "	5 "	(used old dam)	225
17 "	15 "	200 feet	700

The costs are given merely for the power plant and do not include cost of transmission lines from plant to the place where the power was used for electric lighting, etc., nor do they include the cost of wiring the buildings lighted. All this depends, of course, on particular conditions which vary greatly in every installation. These four cases serve to indicate the possible variation in cost of the power plant itself for even approximately the same power. In almost every case an estimate of $50 per horsepower will cover the cost of the plant alone, and from $75 to $100 per horsepower will cover the cost of the entire installation including wiring, lights, motors, etc. The higher the head the less the cost, other things being the same. The larger the plant the less the cost per horsepower usually.

In the table given above, the dam in the eight-horsepower plant was of earth. It cost about $400. In the fifteen-horsepower installation the dam was of concrete and raised the available head 50 per cent. to the figure given. The cost of operation is merely the interest and depreciation on the plant and the taxes amounting to approximately $8 per month. The actual cash outlay for oil and repairs will not exceed $1.25 a month. That is, for less than $10 per month this man has fifteen horsepower available for his service night and day continually. While the first cost is somewhat greater than that of a fifteen-horsepower oil engine, the latter would cost at least $50 per month for continual operation.

The turbine wheel must be installed pretty

close to the dam, but the Pelton wheel is frequently far from the point where the water is available. In the first case the power used is transmitted electrically from the power house at the stream to the place where it is to be used. In the latter case the water is transmitted from the stream through pipes to the Pelton wheel. As a rule there is no dam or similar construction necessary if a Pelton wheel is used. In fact, there need be no running water. A pond elevated above the wheel is ideal. The expense consists of the pipe line to the wheel, the wheel itself, the drain or other arrangement to carry the waste water away. This style of plant lends itself to ready use in connection with irrigation projects. Under these circumstances the water is brought to the wheel, and after leaving it the waste water is led away to the fields to be irrigated. The expense incurred may then be divided between the two projects, power and irrigation, and the total expense for both will but slightly exceed that for either alone.

The cost of the Pelton wheel depends upon the size. Depending upon the head under which it is to operate, a three-foot wheel costs from $220 to $450, a four-foot wheel from $285 to $675, a five-foot wheel from $350 to $625, a six-foot wheel from $400 to $800. The following table gives the horsepower developed by these standard wheels under the various heads. The amount of water needed in each case can be figured by the methods already given.

POWER DEVELOPED BY PELTON WATER-WHEELS

Head in feet	3-ft. wheel h.p.	4-ft. wheel h.p.	5-ft. wheel h.p.	6-ft. wheel h.p.
20	1.5	2.6	4.2	6.0
30	2.8	4.9	7.7	11.0
40	4.2	7.6	11.9	17.0
50	6.0	10.6	16.6	23.9
100	16.8	29.9	46.9	67.4
150	31.0	55.0	86.2	124.0

Smaller wheels may be purchased for smaller heads, or for the same heads and smaller quantities of water than required by these large sizes. Any sized wheel can be used on any head, but with a certain head and a definite quantity of water, a particular size of wheel is best adapted for the development of the greatest power.

CHAPTER XIX

The Storage Battery for the Farm

THERE are only two types of storage batteries which can be considered as valuable from a commercial standpoint, but there are many different storage batteries which will work. A storage battery is merely any kind of an electric battery which, when charged, will be changed from its normal condition into a condition that it cannot maintain. Just as a box will stand up on one corner as long as you steady it with your hand, so will the storage battery remain in its unstable condition so long as it is being charged. When the charging current ceases, the battery tends to return to its original state, and in doing so it gives up a portion of the electricity which was used in changing its condition during charging.

The storage battery does not store up anything in the sense that we can store up water in a reservoir. We may compare the action to that of the water in a steam boiler. When the fire under the boiler acts on the water, it changes the water into the unstable form of steam. The water is charged with energy by the heat of the fire and the steam is merely the charged water. Now, if the fire is withdrawn from the boiler, the steam will condense back to the original form of water. Not all of the heat which came from the fire will be given up by the steam when it condenses, because much of the fire's heat was lost in radiation from the hot boiler containing the water and steam and much of it was lost up the chimney. This will never be regained.

In the same way, the electric storage battery will not return as much electricity as was used to charge it. One type, the lead-plate battery, will return about 80 per cent. The nickel-iron cell of Edison will return only about 60 per cent. In other words, if you use a storage battery for your farm lighting plant and a certain number of charges cost you one dollar, the electricity you get from the battery on discharge is only 80 cents' worth in one case and 60 cents' worth in the other. The remainder of the dollar is lost in operating the battery.

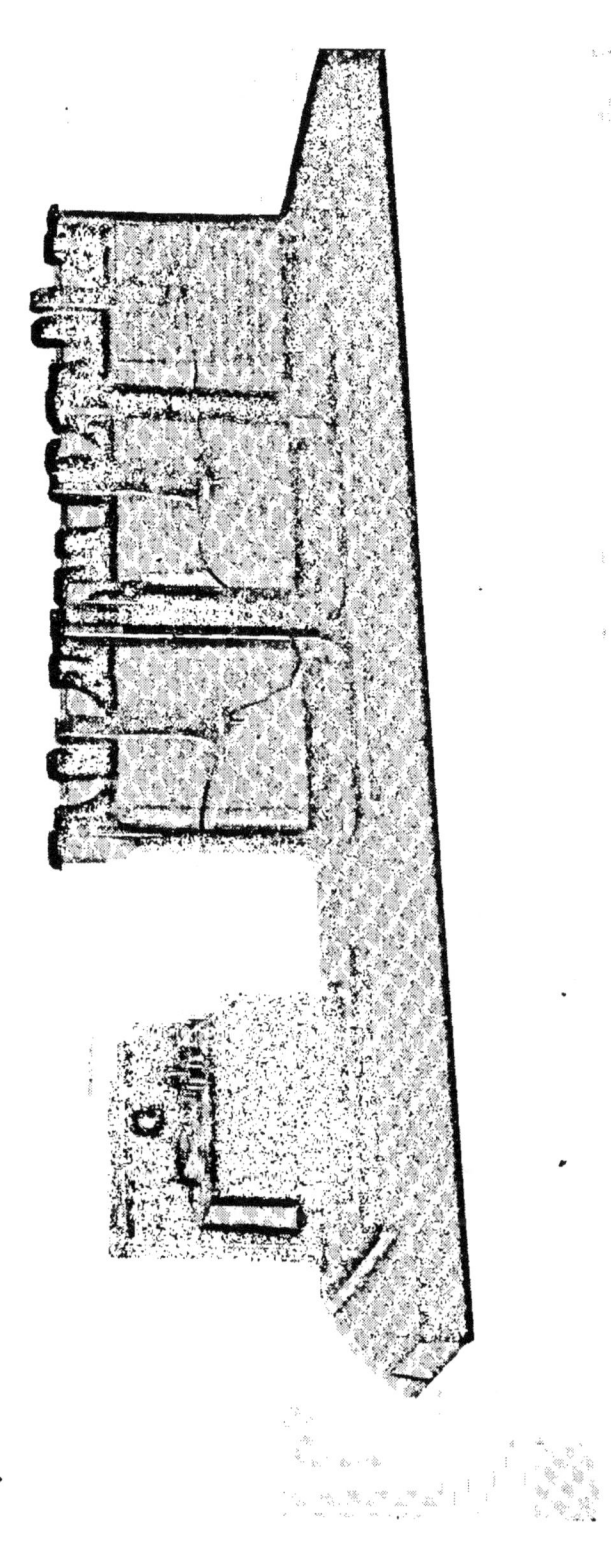

Fig. 32.—Charging a small storage battery with alternating current lighting circuit by means of the current rectifier seen at the left

Fig. 33.

Fig. 34

Fig. 35.

Fig. 36.

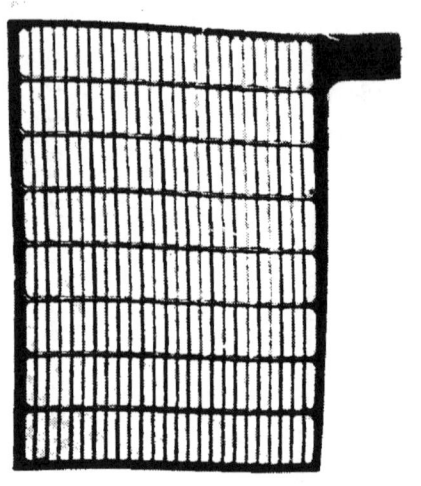

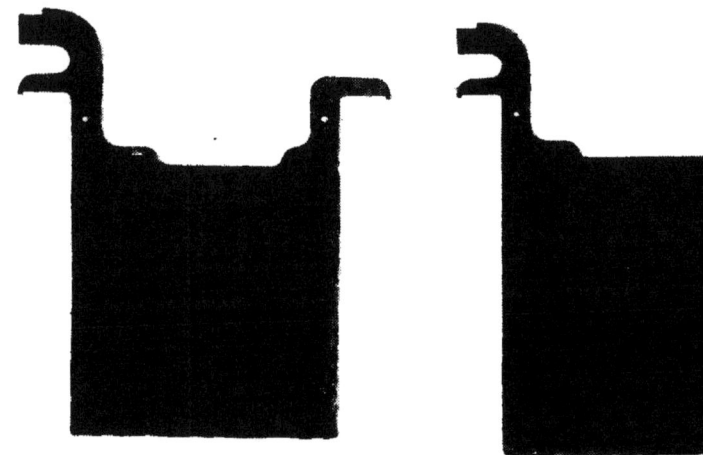

Fig. 33.—The grid before pasting

Fig. 34.—The pasted plate completed

Fig. 35.—The Planté plate after shredding but before "forming" the paste

Fig. 36.—The Planté plate completed and "formed"

The lead-plate storage battery consists of a container, usually glass but sometimes rubber, a liquid chemical called sulphuric acid, and two plates holding a quantity of lead paste. The plates may be of two kinds. One variety, known as pasted plates, consists of a grid or framework made of an alloy of lead and antimony with the spaces filled by the lead paste. One plate, the positive, has a paste of red lead and sulphuric acid, and it is red in colour or, perhaps, a reddish brown. The other plate, the negative, has a paste of litharge and sulphuric acid and when bought is usually gray.

The second variety is known as the Planté type. Plates of this type consist of lead sheets first cut up or shredded and then treated with acid over and over again to form the required paste right out of the lead on the plate itself. This is a long process and makes a more durable but a heavier battery. It costs more than a battery using pasted plates and is more desirable. The chief trouble with the lighter pasted type is that the framework expands and lets the paste fall out, thus ruining the battery. For automobiles and trucks where weight is important the lighter but otherwise less desirable pasted plates are sometimes used.

The Edison battery differs in every way from the lead cell. The container in place of being glass is nickel-plated steel. The plates in place of being the easily injured lead paste arrangements exposed to all kinds of abuse are nickel hydrate and iron oxide packed in strong steel tubes or boxes. The electrolyte or liquid in place of being sulphuric acid is caustic soda.

The particular advantage which is claimed for the Edison cell is its lower weight for the same capacity. Its bulk or size does not vary greatly from similar lead cells, and its efficiency, as already stated, is considerably lower. The voltage or electrical pressure is but 1.2 volts as against 2.1 volts for the lead cell, so that for work requiring a certain voltage as, say 110 volts, nearly twice as many Edison cells will be required, and each Edison cell costs as much as the best type of lead cell. If less weight were its only strong point it would not be in as great demand as it is for other purposes besides electric vehicles. The fact is that its mechanical strength and dependability are enormously superior to the lead cell. It may be abused in every way short of absolute purposeful destruction with but slight damage resulting. This is the feature which commends it for most farm purposes where it must be handled by men who have not been technically trained. It is truly "as rugged as a battleship."

Whether or not a storage battery should be used in connection with a lighting plant depends largely on the purse of the individual. A lighting plant including a storage battery must also include an engine and a generator to charge the battery, besides a switchboard and meters. Such a plant, even of small size, will cost $300 or more, and its efficiency will be very low. A considerable decrease in first cost and an increase in efficiency will be obtained if the lighting circuits are connected directly to the generator, no storage battery being used. In this case, of course, the engine must run all the time lights are required. This in some instances will be most inconvenient unless there is a handy waterfall or a small stream to operate the generator, but the saving of $100 or $150 first cost in doing away with the storage battery and switchboard, the saving of many dollars per year cost of upkeep of the battery, the saving of its depreciation, the interest on the money which would have been invested in it, and the saving of part of the 20 per cent. loss on each charge, will appeal to many farmers as being very desirable even at the expense of some inconvenience.

The capacity and number of cells to be purchased depend upon the voltage of the system installed and the amount of lighting that is to be done. The voltage of the individual lead cell is 2.1 when charged, but it drops gradually to 1.8 volts, below which it must not go. The working voltage is generally figured at 2. The Edison cell, if the discharge rate is normal, starts at about 1.5 volts when fully charged and drops very fast to slightly over 1.2 when the fall to 0.9 volts is gradual. It must not go below the latter figure. The normal voltage is 1.2.

The capacity of a cell is measured in ampere-hours. One ampere-hour capacity means that the cell will give a current of one ampere for one hour. The cell will probably actually give less than one ampere if discharged in as short a time as one hour, because the rating is always at the eight-hour rate of discharge. That is, the bat-

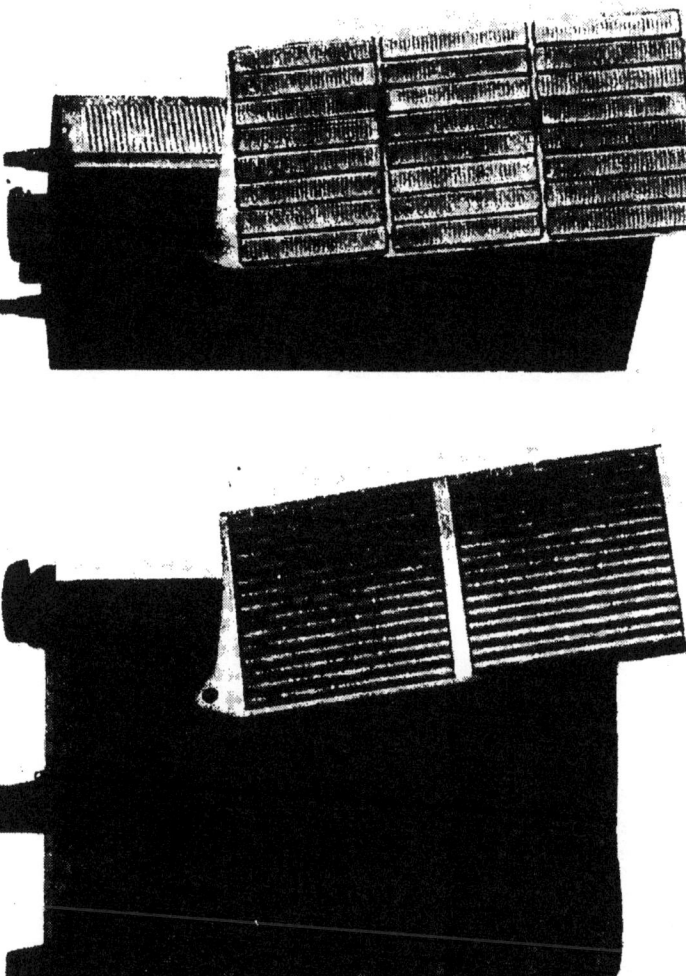

Fig. 37.—The lead cell at the left. The Edison at the right. This view shows Edison plate

Fig. 38.

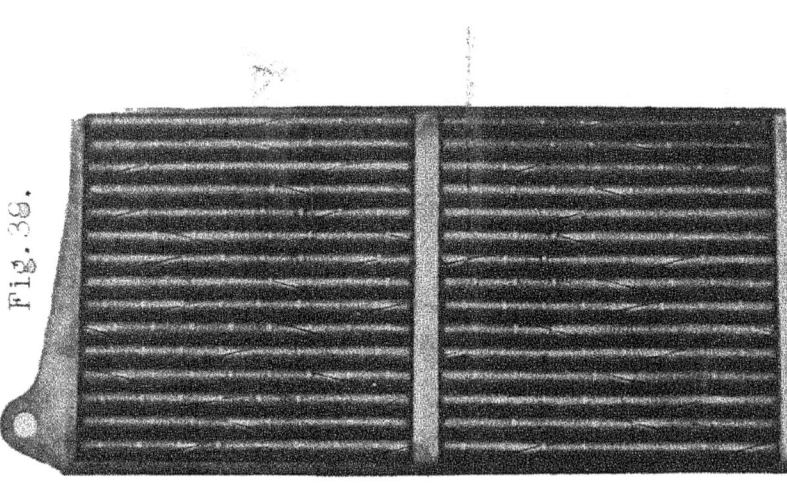

Fig. 38.—The Edison positive plate

Fig. 39.

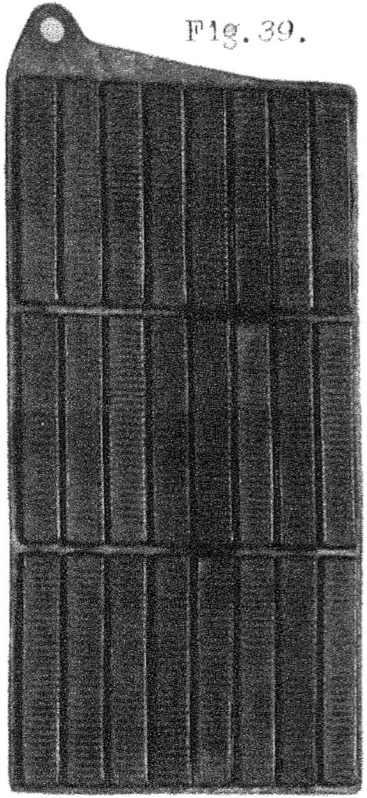

Fig. 39.—The Edison negative plate

tery is rated to give one eighth of an ampere for eight hours, but if rushed to discharge in less time it will not give quite its full capacity. The eight-hour rate is the normal rate.

A 16-candlepower 20-watt lamp requires two thirds of an ampere at 30 volts, the common house voltage for these electric plants. At 60 volts it requires only half the current to operate a 20-watt lamp. The "watt" is the electrical unit of energy. It is the product of one ampere and one volt so, given the watt rating of the lamp and the voltage of the system, divide the watts by the volts and you get the amperes required for a lamp. Then multiply the number of lamps you wish to use at one time by the number of hours they must be lighted, and multiply this by the number of amperes for one lamp under the voltage of the system you use and you get the ampere-hour capacity required for your storage battery. Some allowance should be made for emergencies which may arise requiring the use of more lamps than usual.

The following table gives approximate ideas of costs and capacities of some of the battery systems:

TABLE I

STORAGE BATTERY SYSTEMS

Voltage of system	Number of cells required		Approximate cost of batteries		Capacity in 16 c.p. tungsten lamps for 5 hours	
	Lead	Edison	Lead	Edison	Lead	Edison
6	3	5	$ 20.70	$ 30.00	2	2
30	16	27	110.40	162.00	10	12
60	32	53	220.80	318.00	21	24
100	60	100	414.00	600.00	38	44

As this table indicates, the cost of any storage battery installation depends largely on the voltage of the system used. There must be as many cells as the number given by dividing the voltage of the system by the voltage of the individual cell used. The voltages given in the table are those usually used in small, isolated country plants, the thirty-volt system being very popular. With the higher voltage impressed on lamps of the same rating in watts the current consumed is less, of course, than with the lower voltage, so there may be more lamps lighted even using the same size of storage cell. The following tables afford opportunity for determining the dimensions of the cells and their electrical characteristics:

STORAGE BATTERY FOR THE FARM 153

TABLE II

SINGLE EDISON CELLS

	B-2	B-4	B-6	A-4	A-6	A-8	A-10	A-12
Weight in pounds	4.6	7.4	10.5	13.5	19.2	27.5	34.0	41.0
Normal amp.-hour capacity........	42	84	126	168	252	336	420	504
Amps. discharge at 5-hour rate....	8	16	24	30	45	60	75	90
Charging rate for 7 hours........	8	16	24	30	45	60	75	90
Voltage..........	1.2	1.2	1.2	1.2	1.2	1.2	1.2	1.2
Length..........	$1\frac{1}{2}''$	$2\frac{5}{8}''$	$3\frac{13}{16}''$	$2\frac{11}{16}''$	$3\frac{13}{16}''$	$5\frac{1}{16}''$	$6\frac{3}{16}''$	$7\frac{3}{8}''$
Width...........	$5\frac{1}{8}''$	$5\frac{1}{8}''$	$5\frac{1}{8}''$	$5\frac{1}{8}''$	$5\frac{1}{8}''$	$5\frac{1}{4}''$	$5\frac{1}{2}''$	$5\frac{1}{2}''$
Height...........	$8\frac{3}{4}''$	$8\frac{3}{4}''$	$8\frac{7}{8}''$	$13\frac{7}{16}''$	$13\frac{7}{16}''$	$14''$	$14''$	$14\frac{5}{8}''$
Price with tray...	$6.00	$8.00	$11.50	$13.50	$20.00	$26.00	$33.00	$39.00

TABLE III
A TYPICAL LEAD PLATE-BATTERY

		A	B	C	D	E
Weight in pounds		35	55	72	88	103
Normal amp.-hour capacity		20	40	60	80	100
Amps. discharge at 5-hour rate		3.5	7	10.5	14	17.5
Voltage		2.0	2.0	2.0	2.0	2.0
Length Jar	Glass	$7\tfrac{7}{8}''$	$7\tfrac{7}{8}''$	$7\tfrac{7}{8}''$	$9''$	$8''$
	Rubber	$6\tfrac{3}{4}''$	$6\tfrac{3}{4}''$	$6\tfrac{3}{4}''$	$8\tfrac{1}{2}''$	$6\tfrac{3}{4}''$
Width Jar	Glass	$3\tfrac{1}{4}''$	$4\tfrac{3}{4}''$	$6\tfrac{3}{8}''$	$5\tfrac{1}{4}''$	$9\tfrac{1}{8}''$
	Rubber	$1\tfrac{13}{16}''$	$3\tfrac{1}{4}''$	$4\tfrac{11}{16}''$	$3\tfrac{1}{4}''$	$7\tfrac{9}{16}''$
Height Jar	Glass	$17''$	$17''$	$17''$	$20\tfrac{1}{4}''$	$17''$
	Rubber	$13\tfrac{3}{4}''$	$13\tfrac{3}{4}''$	$13\tfrac{3}{4}''$	$16''$	$13\tfrac{3}{4}''$
Price Jar	Glass	$4.34	$6.90	$9.37	$9.94	$14.12
	Rubber	6.10	9.16	12.22	12.40	18.35

Table III gives only one make of cell and only one line except cell D, which is another line altogether, yet it is made by the same manufacturer as the others. This cell, although approximately the same price as cell C, has a third greater capacity. This is owing to the fact that the plates are larger and thus cost less to make per square foot of their surface, and the size of the plate surface determines the capacity of the battery. Whenever batteries are purchased it is important to state to the

Fig. 40.—A kerosene engine belted to a lighting generator

manufacturer not only the capacity you require and the number of cells, but also the amount of space you have available for storage. Some manufacturers will furnish lead-lined wooden tanks with properly arranged compartments for the entire battery in place of the fragile glass or expensive rubber receptacles for the individual cells. In some instances this will save money on the initial outlay as well as upon the repair costs.

PART IV

DRAINAGE AND IRRIGATION

The Principles of Drainage.
The Construction of a Tile Drain.
Some Facts Concerning Small Irrigation **Practice.**

CHAPTER XX

The Principles of Drainage

The immediate purpose of drainage is evident to any one. It is to remove the surplus water from the land. The reasons why this improvement results in better crops is not so obvious. It is the purpose of this chapter to explain briefly what drainage is, the reasons for drainage, the desirable drainage methods and principles underlying them, and to give some practical information about drainage systems, thus making clear why better crops result from carefully planned artificial drainage. The next chapter will explain in detail how to tackle any particular job of tile drainage.

First, it is necessary to call attention to the well-known but easily forgotten fact that the soil everywhere is permeated with moisture. All soil has some water in it even when it appears to be quite dry. This water may be of two kinds, either hydrostatic (ground water) or capillary moisture. The latter fills the small pores between the particles of soil above the ground water level. The former fills all the spaces, big or little, below that level. The ground water level is, of course, a variable thing, constantly shifting as the seasons are wet or dry. Its location at any particular time may be determined by digging a post hole or well in the soil. The level at which the water stands is the ground water level at that point. In wet weather the hole may be pretty full, while in dry weather the water level may be several feet below the surface. For crops to thrive and prosper the ground water level must be several feet below the surface most of the time, and capillary attraction, such as exerted by any porous body as a lump of sugar or a sponge, is depended upon to lift or suck up from the ground water sufficient moisture for the use of plants.

In clayey soils the water does not drain off readily after a storm. Puddles and mudholes form on the surface and only dry up by evaporation. The ground water level is then high, being lowered gradually by the slight seepage through the soil and by evaporation from the surface. Before it lowers sufficiently, however, another storm comes and raises the level again. Thus the ground water in such soils is always high except in times of extreme drought. It is such soils that are benefited by drainage.

Any soil where the natural drainage is poor so that the ground water level is less than about three feet below the surface at plowing time will be benefited by artificial drainage. Of course all soils must be drained in order that they may be tillable, but in most cases this is done naturally by seepage through porous layers to an impervious layer where there is an underflow of ground water carrying off all surplus moisture. All soils which are sandy or porous at the surface may not be well drained naturally, however. There may be a clayey subsoil close to the surface which is nearly as detrimental to good drainage as though it came entirely up to the surface. Occasionally such lands cannot be drained advantageously. In other cases the surface soil may not be as porous as expected, but the subsoil may be very open and give perfect underdrainage. It is not possible, therefore, to tell from mere inspection of the surface soil whether or not artificial drainage is needed. In general, what is known as a "wet" soil needs drainage as does also a "late" soil, while a "dry" soil or an "early" soil is well drained naturally.

The reasons for drainage are many, but all are contributory to one great reason, and that is to raise a better crop. In many cases crops have been more than doubled on land which was tillable before, while the reclamation of swamp land and production thereon of a splendid crop is extremely common. Drainage opens up the soil by removing the surplus water and allowing the air to enter. The air currents circulate all through these porous soil layers. This is beneficial in many ways. It makes the soil more friable and less likely to cake. It assists necessary chemical actions in the soil. It promotes the growth of bacteria which are necessary in order that the soil materials may be changed into plant food. Air is essential to rugged root development. If the ground is water soaked so that the roots must cling close to the surface to breathe, then in times of drought the water level goes so far below them that capillary attraction will not raise water for their use. If, on the other hand, the upper soil contains free air, the roots strike deep into the soil and the comparatively slight change in water level at drought times does not affect them. They have a larger area of capillary tubes to bring them their water supply.

It is a fact that drained soil when workable has far more water in it than undrained soil when in the same tillable condition. Well-drained soils are more open and less compact and therefore hold more water in suspension than undrained soils. They have greater capillary power because they are more porous, so that long after crops in undrained soils have perished those tiny capillary tubes in the porous soil supply water to their plants. The small amount of water that does fall in the dry season can be much better conserved in drained soils because it is possible to get on the land at once and cultivate it.

Drained soils are from 5 to 10 degrees warmer than undrained soils in the spring months, and somewhat warmer most of the time. It is possible to get on a drained field from three weeks to a month earlier in the spring because of this. The water is carried off and the air gets in the soil, warming it up. There is not the cooling effect of continued evaporation. The sun's heat warms the soil and doesn't have to first evaporate the surplus water. Seeds germinate better in a warmer soil, and by getting such an early start the crops are well along in hot weather so that they can stand drought better, besides allowing the farmer finally to take advantage of the early markets. To a great extent the "freezing out" and "heaving" of winter grain and of posts will be prevented if the soil is drained, because the water will be drawn downward and not allowed to freeze. Drainage prevents washing and floods in wet growing seasons. It is, in short, a stabilizer which makes uniform growing seasons, allowing the growth of a good crop every year whether it is a wet season or a dry one.

Underdrainage is valuable for fertilizing purposes. It has been mentioned that the passage of air freely through the soil helps the chemical and bacterial actions which occur. Fertility is also added to the soil with each fall of rain or snow, for as the water is drawn through the soil it is deprived of the nitrogen and carbonic acid which it took from the air in passing. This is all lost if the water is carried away by a surface flow. Manure and other fertilizer are drawn down through the soil layers by the water, and surface washing is prevented.

Surface drainage by ditches will not give all these benefits but it is beneficial in many places, for it lowers the ground water level and permits cultivation. Underdrainage is, however, the advisable method, and this can be best accomplished by modern tile drains. Both cement and clay are used for tiling, and there really is not much choice between them if they are properly made. Usually cement tile, if used, is made at home by the farmer while clay tile is purchased. Under these conditions the cost of the cement tile is about half that of the clay, if labour is not counted. The mixture used is one part cement to four parts of clean, sharp sand for the smallest sizes, and one part cement to three parts of sand for the larger. With a hand machine two men can construct 1,000 feet of four-inch tile in about two days of eight hours each at a cost of from $8 to $10 for the materials used. The cost of clay tiles is approximated in the following table, including freight rates for 100 miles. As this tile is made in every section of the United States, no difficulty should be met with in getting it at any time and in any quantity.

PRICES, WEIGHTS, AND COST OF LAYING CLAY TILE

Tile Diameter	Price per 1,000 ft.	Cost per rod for laying 3 ft. deep	Pounds per ft.	Average carload
4 in.	$ 18	$0.33	6	6,500 ft.
5 "	26	.33	8	5,000 "
6 "	35	.33	11	4,000 "
7 "	45	.35	14	3,000 "
8 "	60	.40	18	2,400 "
10 "	80	.45	25	1,600 "
12 "	120	.50	33	1,000 "

The grading, depth, and spacing of tile for any particular job depends upon local conditions and on the size of tile used. Large tile, as used in mains, may have a fall as slight as one inch in 100 feet, but small tile used for laterals or side branches must have at least twice that amount. A greater fall is desirable. Lines of tiling across a slope to prevent the seepage of water down the slope should have considerable fall; six to ten inches in 100 feet is little enough.

The depth of lines of tile depends largely on the spacing of the lines. The deeper they are the farther apart they may be spaced. The closer the lines are laid, the shallower they may be placed. Three feet is the common depth in clay and four feet in a sandy soil. In the

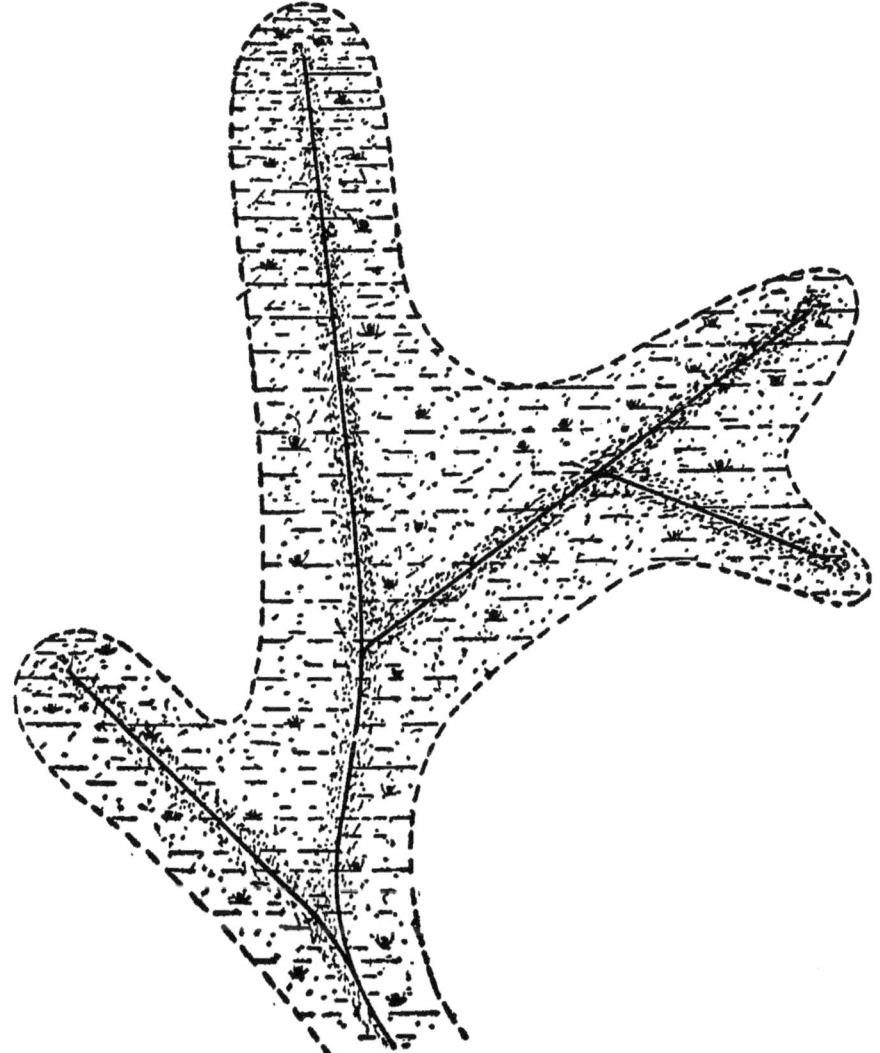

Fig. 41.—A natural system of drainage

former case the laterals are placed four rods apart; in the latter, eight rods apart. Portions of the field may need closer lines. A simple calculation shows that in the former case forty rods of tiling is needed per acre for the laterals, while if, as in the latter case, they are eight rods apart, twenty rods of tiling is enough.

The cost of draining depends, of course, on the depth and spacing. It ranges from $14 to $40 per acre. The average cost in a great number of cases was $25 per acre, using four-inch tile for laterals and six to ten inch for mains. The following table shows the approximate cost per acre under various conditions, using a four-inch tile at $18 per thousand and estimating cost of laying as given in the previous table. This table following makes allowance for digging ditch, purchasing tile, laying tile, and refilling ditch. Cartage will have to be added:

APPROXIMATE COST OF TILING PER ACRE

Spacing	Feet required per acre	Cost per acre
200 ft. apart	218 ft.	$10
150 " "	290 "	13
120 " "	363 "	16
100 " "	436 "	18
80 " "	545 "	24
60 " "	726 "	32
50 " "	872 "	38
40 " "	1,090 "	48
30 " "	1,450 "	64

The systems of tiling used are five in number, the natural system, the gridiron, the herring-bone, the single line, and the cross-the-slope system, all of which are shown in the accompanying diagrams. In the natural system single lines of tile are placed along the lowest part of the various wet marshes, no attempt being made to lay out a systematic design. It is, in many ways, the most economical plan and frequently the most efficient. The gridiron system is used in flat fields requiring thorough covering. It is very economical. The herring-bone is rather a cross between the former two. It is used in broad fields with a natural depression running through it, along which the main is placed. It is not so economical as the gridiron plan but is necessary in some places. The single line is used for inexpensive first-cost systems. Its upkeep may be somewhat greater than the others because each line has an outlet which must be kept free and clean. The cross-the-slope arrangement is used to intercept water flowing down the hill in rather moist hillside areas having considerable slope. The choice of the system to be used depends on the individual. In most cases the gridiron arrangement is desirable. In every case the main should run along the lowest ground and the laterals run parallel with each other and

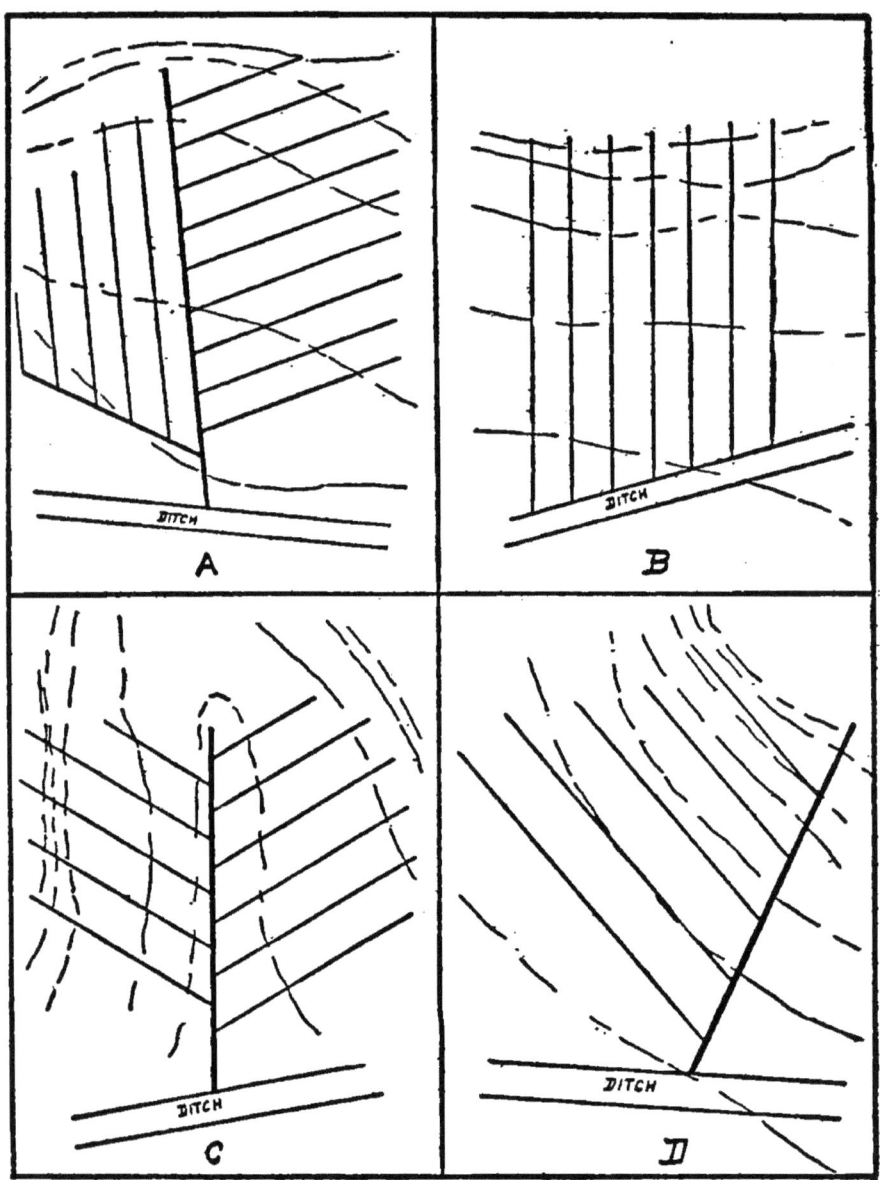

Fig. 42.—Other drainage systems. *A*—Gridiron. *B*—Single line. *C*—Herringbone. *D*—Cross-the-slope

with the slope, unless the latter is very steep. It is thus seen that the field will, to some extent, lay itself out and choose its own system.

Frequently catch basins are desirable along the line of drains to carry away some particular flow of surface water. They are easily placed at the time the drains are laid. In every case care should be taken to prevent dirt and silt entering the drain at these places. Such tile lines should be six inches in diameter at least. The basins are merely pits filled with broken stone. Occasionally they are walled-up cisterns built of concrete covered with a protective grating.

The profits of draining depend largely upon the skill shown in planning and executing the work. Not always will the first crop pay for the drainage installation, but in every case where intelligent management is used four crops should show such an increase as to more than pay for the system and interest on the investment during that period of time. In hundreds of cases one crop on land which had never before felt the plow has paid many times over for the drainage operations. In a few cases drainage has not increased the crop any but has made it earlier and so of more value.

CHAPTER XXI

THE CONSTRUCTION OF A TILE DRAIN

THERE are at least four distinct steps in the construction of any drain: (1) locating the outlets and laying out the tile lines; (2) digging the ditch; (3) making the outlet bulkheads and laying the tile; (4) filling the ditch. The first step includes making a sketch, however rough it may be, of the land to be drained. Note on it the slope of the land and the points of the compass. Mark the line of lowest land. Make this sketch complete for a whole drainage system, no matter how little is to be done at the present moment. Then the little that is done may be made to accord with the plan for the whole, and trouble later on may be avoided. Then with this sketch in hand go over the land carefully and follow the general order given below. This is suggestive and not exacting, but it has been found from much experience to give the most desirable results.

1. Locate the outlets of the mains at the lowest points of the land emptying into a present waterway or into an open ditch leading to some waterway. Have as few outlets as possible, for they are always troublesome and must be carefully watched to prevent clogging. Usually one is enough. It will pay in most cases to build a concrete casing for the outlet, and across the opening embed bars of iron to prevent the entrance of anything.

2. Locate the main or mains along the line of lowest ground. This can be determined by following the channel along which the surface water flows in flood times. If the field is nearly flat, locate the main along one side, as by so doing you can save tiling and save joints.

3. Mark line of main and lines for laterals by driving stakes firmly into the ground every fifty feet. Make the laterals parallel to each other and, so far as possible, have all lines straight. If curves are necessary make them with a long radius. Avoid sharp bends.

4. Determine total fall of each main and lateral by using a sight level. Divide by the number of stakes and thus get fall between each two consecutive stakes.

5. Calculate depth of ditch below top of each stake to give proper fall. Make depth gauge of a pole and a cross-piece at this calculated depth away from the end of the pole.

6. Stretch a cord from top of one stake to the next as a guide in digging. This line should have the proper calculated fall which the ditch is to have, so that if the depth gauge is placed anywhere with the cross-piece on the line the end of the pole will be on the bottom of the ditch.

7. Begin digging at the outlet or, if outlet is to empty into an open ditch which runs to some waterway, dig this ditch first. Then begin the tile drain main ditch at outlet, making ditch a foot to one side of line of stakes so that position of stakes will be preserved. Start ditch with plow and then use hand tools. Dig ditch just wide enough to stand in, perhaps fourteen inches wide at top and eight inches at bottom for a ditch three feet deep.

8. After getting ditch the right depth at each stake, sight along the bottom to remove all humps and hollows and make a smooth, gradual, and continual slope. Test this grade in many places by measuring down from line with the

depth gauge. This part of the work is very important and needs care.

9. Groove the bottom of the ditch carefully and lay the tile in the grooves. These grooves are of great value in laying round tile. The tile should be laid as fast as the ditch is dug and put in shape. Start laying the tile at the outlet.

10. Begin at outlet and lay main. The first few feet should be of glazed or hard-burned tile to resist frost. The first few joints should be cemented.

11. Wall up outlet with stone to protect it, or build the concrete bulkheads. This is important and should not be slighted. It may be postponed until later, if desired, but must not be forgotten.

12. Lay main as far as first lateral and put in a Y-connection and lay the first few feet of that lateral. Keep on laying main and the adjacent ends of the laterals until main is complete.

13. Go back and finish laying laterals.

14. Place a flat stone against the upper end of each tile line so as to close it against the entrance of dirt or rubbish.

15. Cover every hole or crack in or between

tile larger than one quarter inch by a piece of broken tiling.

16. Cover tile with loose earth carefully. Then fill ditch in most convenient and rapid way. A scraper made of a plank on edge something like a snow-plow will do good work if pulled along the ground at one side of the ditch. So will a road scraper or a split-log drag. By turning several furrows into the ditch with an ordinary horse-plow, the filling will be very rapidly done.

The following suggestions may be of value in connection with the work:

1. Plan the work and start deep enough to drain the whole field.

2. If the natural slope is not good enough the ditch may be made a little shallow at the upper end, grading to the proper depth gradually. Remember that the grade should be uniform along any length of tile.

3. Get all the fall possible.

4 Make joints real tight. Water can get in where sand and dirt cannot. No opening should be more than one fourth of an inch.

5. In quicksand cover each joint with a

piece of roofing paper. A little concrete laid in the bottom of the ditch and grooved will help greatly in maintaining the grade through a run of quicksand.

6. Keep an accurate map of the whole field, marking carefully the location of joints and dead ends.

7. Cover end of last tile laid if work is interrupted so as to prevent filling if a heavy rain comes on before work is resumed.

8. The first dozen joints next to outlet should be cemented tight to insure perfect results. The greatest care must be taken of outlets.

9. Use no tile smaller than four inches in diameter. It doesn't pay. They fill up quickly and their first cost is but slightly less than the four-inch. Remember the four-inch has nearly twice the carrying capacity of the three-inch, and variations in grade which would be disastrous to the three-inch will not greatly affect the larger sizes.

10. Every tile should be perfectly round and should give a clear, metallic ring when struck.

11. Open ditches should be seven feet deep and at least three feet wide with 45 degree sloping sides.

CHAPTER XXII

Some Facts Concerning Small Irrigation Practice

It is now recognized that practically all crops may be benefited by proper irrigation where water is cheap and plentiful. It is not as universally known that proper drainage is essential to make the benefits from irrigation as great as possible. The danger without drainage is that the raising of the ground water with consequent capillary rise and evaporation will cause too great an accumulation of undesirable soil salts in the surface layers of the earth. This is a subject that has attracted the attention of many experts and what is referred to when the statement is made that continued irrigation is the cause of soil deterioration. Proper cultivation of irrigated lands and care in the use of water will do much to offset the disadvantage of poor drainage. Cultivation of the soil after applying the water will prevent rapid evaporation and will allow the crops the full use of the water applied, thus making for economy in water.

The desirability of cultivation leads to the belief that the method known as subirrigation is the best one to follow. It has received much thought and study, but the results from it are entirely unsatisfactory because of the initial outlay involved and the fact that for many crops the inequalities of distribution are fatal. The furrow system is, on the other hand, the cheapest, simplest, and probably the most widely used method. Lately, too, a method of sprinkling has been used with success on small fields, known in some sections as the "Skinner Irrigation System."

Particularly in sloping fields is the furrow system easily laid out. The furrows are run down the slope, either directly or diagonally, on an angle depending upon the amount of grade. Sometimes they are laid out in wide, sweeping curves. The steeper the grade the nearer to the horizontal must the furrows be cut, the grade being thus lessened. Such furrows are connected to the main furrow by curved flumes. The main feeding furrow runs along the top of the grade at the upper ends of the laterals.

More than one main or flume will be needed in most cases, these being spaced apart down the grade an amount depending upon the distance a stream will run in the branch furrows. No rule can be given for this, as it depends entirely upon how much water is flowing, that is, upon the size of the stream in the furrow, and also upon the character of the soil. In porous soils furrows from 40 feet to 200 feet will be as long as desirable, while in closer packed land furrows may run as long as 600 feet. Usually they are from one foot to four feet apart and are from three to twelve inches deep. The deeper they are made the less evaporation from them per unit in the same length of time.

Although particularly adapted for use with crops planted in rows, the furrow system is of great value in all types of planting. It is very adaptable and may be widely modified. By damming the furrows at any time the fields may be readily flooded, if desired, and the advantages of the flooding system obtained. This is very desirable with extremely porous tracts where the water must be gotten over the ground quickly. The advantages of the furrow system are that the entire surface of the ground is not wet and therefore is not in a condition to bake; there is less evaporation than would be with a thin sheet spread out over the field; the water is near the roots and a deeper root growth is promoted.

The main ditches for irrigation projects should, if possible, be laid along boundary lines or fences in order that they may not interfere with other operations. Often it is desirable to convey the water from its source to the furrows by means of concrete or clay pipes rather than with open ditches. It cuts up the land less to do this and is less likely to hinder farm work in general.

To insure equal distribution of water in furrows, boards may be placed in the banks of the main ditches and adjusted so that the proper amount of water flows over the edge. The time taken for the water to run through the furrows may be between fifteen minutes and two hours, depending upon a multitude of conditions. The flow per furrow ranges from fifteen to thirty gallons per minute. The irrigation flow usually continues for perhaps two or three days with an interval between applications of from seven to twenty days.

The Skinner system requires an elevated tank or a pump connected to a water course and able to keep up a continuous supply for the required length of time. The main sprinkler pipes are usually not over 250 feet long but there may be a number of them. Every three or four feet there are outlets or faucets. The pipes for lengths such as this are two inches in diameter and the outlets are three fourths of an inch. A supply which will provide about fifty pounds pressure is satisfactory for a system of this kind, and there are several working well under somewhat less pressure. A water supply of 1,000 barrels will supply an acre and a half with sufficient moisture for about four days during the dry season. Obviously this system is of greatest value in small plots, and the operation of the various sections of pipe may be regulated to suit the needs of that particular place.

In every case it must be remembered that irrigation which provides continuous moisture is better than one soaking and then a dry spell followed by another soaking. Too slight an amount of water should not be applied, but, unless the soil is porous and able to retain large quantities of capillary moisture, a real heavy application should be avoided as should also an extended period between applications. A medium amount applied quite frequently is desirable to promote the continual and rugged growth of the plants. The water should be applied each time before the plant shows signs of distress in order that there may be no hesitation in growth and no tendency to put out undesired new shoots when growth is again fostered by moisture applied after a dry period. The plant and the soil should be watched together, frequent and careful investigations being made of their condition.

The cost per acre to put any field in condition for irrigation varies widely with the system used and with the local conditions. It ranges between $5 and $20. The cost of applying the water afterward is also variable, as might be expected. It is usually figured either as the cost of applying an amount of water equivalent to a foot deep over an acre (called an acre-foot) or as the cost of irrigating an acre for a season or a year. The following tables give the amount of water required to cover an acre to any certain depth:

FLOW OF WATER REQUIRED TO COVER AN ACRE

Equivalent depth of water over the acre	Gallons required to cover the acre	Gallons per minute required to furnish this amount in 24 hours
1 inch	27,157 gallons	18.9 gals. per min.
2 "	54,314 "	37.7 " " "
3 "	81,470 "	56.6 " " "
4 "	108,627 "	75.4 " " "
5 "	135,784 "	94.3 " " "
10 "	271,567 "	188.6 " " "
12 "	325,880 "	226.3 " " "

The next table gives the same information but in a little different form, being arranged to show the amount of surface any particular flow will cover to the depth of one inch.

ACRES COVERED BY ANY FLOW PER MINUTE

Gallons per minute flowing for 24 hours	Total gallons used per 24 hours	Number of acres covered by this to depth of 1 inch
10	14,400	.53
50	72,000	2.65
100	144,000	5.3
200	288,000	10.6
300	432,000	15.9
400	576,000	21.2
500	720,000	26.5

In most cases and for most crops a flow of five to six gallons per minute per acre is the required rate. For wetting the land to a depth of four feet, from about 2.5 to 6 inches of water over the land is required, depending upon its degree of moisture when irrigation starts. Many investigations have given the average figure for complete irrigation at from 4 to 6 acre-inches of water per month, equivalent to a continual flow of about 2.5 to 3.8 gallons per minute per acre during the month. Of course such a continuance of flow is never followed, but a larger amount is used for a shorter length of time.

The cost of application depends upon the system used, the amount and cost of water consumed, the frequency of application and labour costs in the locality. In the East the cost of applying a total depth of four to eight inches of water per acre per year, using the furrow system, and with wages approximating $1.50 per day, will range from $25 to $75 per acre. The sprinkler system under the same conditions will result in a somewhat larger cost usually, but in no case is it likely to exceed $100 per acre per year. If the water is city water and purchased, the cost of it alone will probably amount to $50 per acre-foot, while pumped water will cost but $12 or $15 per acre-foot. In the South and West somewhat smaller costs of water prevail, the cost of pumping being considerably less than that stated, and purchased water being obtained in many places for from $15 to $20 per acre-foot. The cost of applying the water per acre-inch per irrigation will run from 50 cents to $1.

PART V

MISCELLANEOUS ENGINEERING TALKS

The Cost of Road Building.
The Working Principles of Orchard Heaters.
The Forms of Electricity.

CHAPTER XXIII

The Cost of Road Building

CONDITIONS vary so much in various states, and even in different counties of the same state, that it is impossible to give an adequate treatment of this question without going into the whole problem thoroughly. The cost of building the road depends upon many factors such as the cost of labour, availability of materials, the kind of road, the width, the depth of surfacing, the cost of bridges and culverts, the traffic expected, and the climate, particularly as regards rainfall and freezing. There are five distinct kinds of road construction used in this country for pikes—earth roads; sand-clay, gravel, macadam and bituminous macadam. In Texas there is also considerably more than a hundred miles of shell roads, and there are similar lengths in other coast states. The following table gives some idea of the wide variation in the cost of roads in different sections.

TABLE OF COST OF ROAD CONSTRUCTION

| | Cost of Roads per mile ||
	In Texas	In U. S.
Earth roads	From $60 to $400. Average in 5 counties, $168.	
Sand-clay roads	From $60 to $2,000. Average in 41 counties is $593.	From $387 to $1,775. Average in 17 states is $723.
Gravel roads	From $100 to $4,000. Average in 27 counties is $1,708.	From $940 to $5,950. Average in 31 states is $2,047.
Macadam roads	From $1,000 to $3,500. Average in 5 counties is $2,160.	From $2,153 to $9,164. Average in 34 states is $4,989.
Bituminous macadam	In El Paso County the cost is $6,000.	From $6,000 to $19,681. Average in 10 states is $10,348.
Shell roads	Average in 3 counties is $3,083.	

In Texas the total mileage of all public roads is about 128,971, of which only 2 per cent. are improved. Of the 2,768 miles of improved roads by far the greatest mileage (about 2,254 miles) is of sand-clay construction.

The variation in the cost of road building in any given way depends on the width and depth of material among other things. For sand-clay roads the average width in seventeen States is seventeen feet and the average depth of surfacing is nine inches. The gravel roads in thirty-one States averaged thirteen feet wide and seven inches deep. The average width of a macadam surface is thirteen feet in thirty-four States and the depth is six inches. The average width of bituminous macadam in ten States is fifteen feet and the depth six inches.

Without doubt the sand-clay road obtained by using six to eight inches of clay plowed and harrowed into a sandy gravel to form a thorough mixture gives a comparatively low-priced and satisfactory road, and it is rapidly growing into great favour.

The kind of road and method of maintenance determines the cost to a great extent. The European countries long ago saw the absurdity of building good roads and neglecting them. They therefore established a patrol system. Only in New York State, however, have we had such a system until recently, and there it has been very successful. One patrolman can care for six to twelve miles of road, patrolling the entire section at least twice a week, filling in ruts and holes, repairing defective spots in the surface, sweeping the water-bound surface, and, in general, keeping the road in first-class condition, and the ditches, culverts, etc., clear so that the road may be well drained. It is the practice of the old adage, "A stitch in time saves nine." By such a careful system the cost of maintenance, based on the pay of patrolmen at $3 per day for five months in the year, is about $75 per mile per year. The present form of maintenance of roads in New Hampshire and elsewhere is by use of oil treatment covered with sand. That costs $440 per mile.

There can be no question but what the automobile is the cause of great deterioration of roads. The surface is loosened by the tangential push of the tire as it grips the road. Then there is much slip of the tire on the road and the rubber picks up much of the small stuff. In a test under unusually heavy traffic of unusually heavy motor vehicles, a stretch of gravel road cost over $2,000 per mile for five months' maintenance as against a negligible sum before the circumstances which caused the heavy motor traffic.

For roads requiring merely occasional scraping and dragging to keep them in good repair, $5 per year per mile is an average figure.

As to the distribution of cost, in New York and a number of other States the trunk lines are built at the cost of the State, while other roads are built at joint expense. Usually the State takes charge of the work and pays 50 per cent. of the cost, the county paying 35 per cent. and the town 15 per cent. The maintenance is also divided up.

CHAPTER XXIV

The Working Principles of Orchard Heaters

Many of the Eastern farmers have found out that orchard heaters are not as satisfactory in one orchard as in another, and wish to know the principles of operation in order that they may be used with the greatest efficiency. Others have tried them but once and without success. Full stories of experiences are hard to get without being coloured somewhat by the prejudice of the writer. This plan of orchard heating in times of unseasonable frosts has been tried out for years in the Western States with great success. Fruit growers in New York, New Jersey, Connecticut, and Massachusetts have taken it up more or less in the last few years. In Canada very few orchards are so protected. The basic idea is to start a multitude of small fires in various parts of the orchard when the temperature goes so low as to give a possibility of injuring the crop, particularly about blossom or bud time. The usual fire is a can of burning oil which gives off a dense smoke. Sometimes soft coal is burned, but it is less satisfactory because of the time required in starting it and the fact that it cannot be readily quenched without dumping and wasting considerable fuel. The oil heaters, on the other hand, smoke up well and burn from the beginning, and, if there is a cover on the container in which the oil is held, these heaters may be put out by simply closing the cover and shutting off the supply of air. As many as three or four thousand of these small cans are used in some of the orchards.

The protection afforded comes largely from the great cloud of smoke which hangs low over the orchard, holding in the heat from the fires. If a strong wind gets at this cloud and dissipates it readily, the heaters will not be satisfactory. If the orchard is located high and unprotected in order to get good air drainage, the chances are that this form of heating will be very difficult to arrange. The best location is one that is somewhat sheltered, as where there has been a windbreak erected or where there is a natural windbreak. Particularly in valleys surrounded by small hills this method of frost fighting is very successful. In such places as these the cold winds are prevented by the smoke clouds

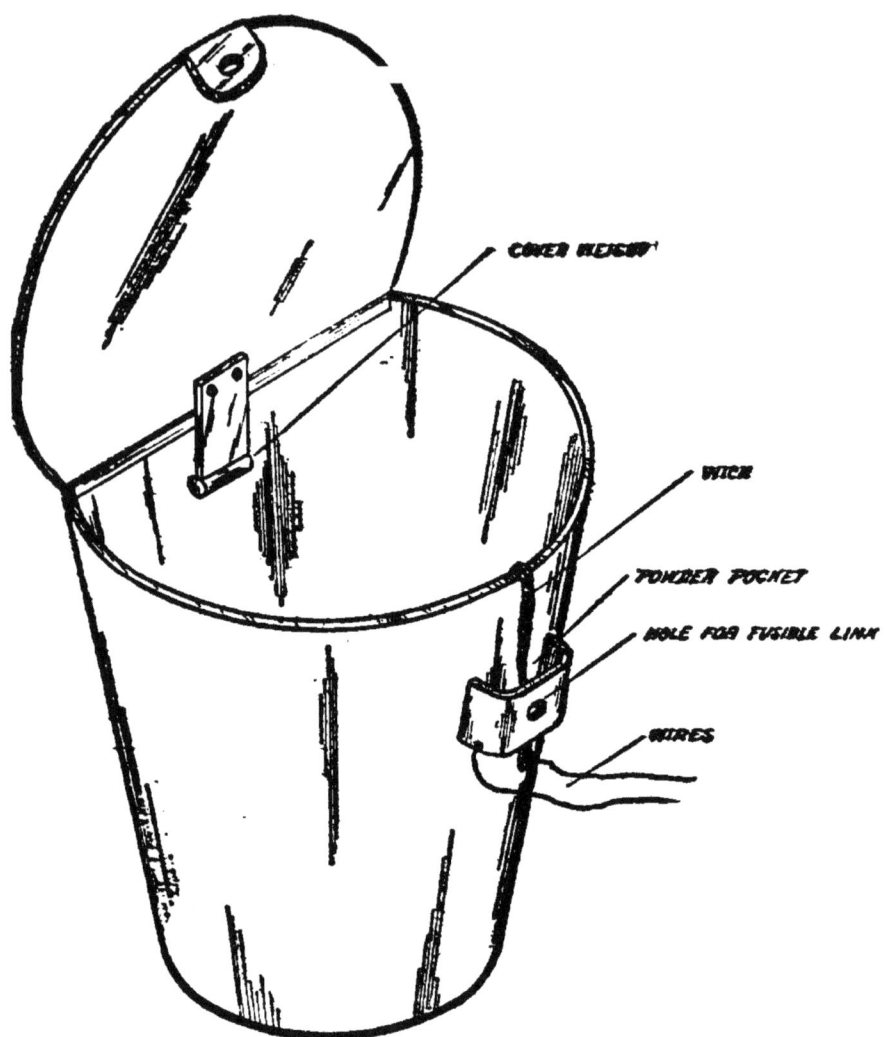

Fig. 43.—An orchard heater

from driving in and making cold air pockets around the trees. Many farmers in setting out new orchards arrange windbreaks against the winds found to be most damaging, with the idea of utilizing the orchard heaters when the trees come into bearing.

The particular type of burner is, of course, immaterial so far as effectiveness goes. It is merely a case of convenience. Small fires of damp brushwood or sawdust, perhaps with a little soft coal thrown on, have been used successfully in the early days of experimentation and are still retained by some fruit growers. One man built his fire on a portable arrangement and dragged it slowly in and out through the orchard with really remarkably good results, but, of course, at the expense of considerable labour and inconvenience. The difficulty of starting such a number of heaters in a short time has been solved by a simple electrical device that any farmer can make. The oil container, built in something the shape of a milk pail, has a hinged cover with a weight attached to it tending to hold the cover open at all times. A piece of fusible metal is tied so as to hold the cover down. In a little pocket alongside of this fusible link there is a small amount of gunpowder and a wick leading from the oil can, as shown in Fig. 43. There is an ordinary gasoline engine spark plug arranged close to the powder. If the plug is too expensive, just the bared ends of two wires held securely a very tiny distance apart will do. These spark gaps are connected in circuit with a spark coil, so that by closing a switch from the battery to the coil a spark will be caused to jump the gaps. In doing this the powder is ignited and lights the wick. The burning wick melts the fuse, releasing the cover and allowing the weights to pull the cover open. The wick thereupon ignites the oil within the can. By setting a stop, the cover may be opened a small or a large amount, thus regulating the fire to suit the conditions.

CHAPTER XXV

The Forms of Electricity

Electricity is the same substance no matter whether it is generated by chemical action, by friction, or by any other one of the many ways in which it is possible to generate it. Just in the same way water is the same substance whether it is in the form of ice, of water, or of steam. Yet, although water consists of a combination of oxygen and hydrogen, and it is the same combination in whatever form the water is, we know that steam will act on a thing differently from what ice will. That is because steam is at a higher temperature than ice, or steam may be under pressure while ice is not. Just so with electricity. When generated by friction, electricity is at high "potential" or pressure, while when generated by chemical action, it is at a low "potential" or pressure. The electricity generated by friction is generally (although incorrectly) called "static" electricity. It is of the same nature as that which causes lightning, the "northern lights," etc., and which is found in the atmosphere. On a dry cold morning as you comb your hair, you hear it crackling because of the discharge of "static" electricity. If you tear up a piece of paper into fine bits and bring them near a glass rod or chimney, or a piece of sealing wax which has been rubbed briskly with fur, silk, or wool, the paper will jump toward the glass, and after a little while, if the paper is in real small pieces, the bits will fly away from the glass. This is all due to "static" electricity. It was the only kind known until about 1792. Then electricity generated chemically was discovered by Galvani. This is known as "galvanic" electricity and is what is obtained from batteries.

Afterward, electric dynamos became known and the electricity obtained in that way is called "dynamic" electricity. All these names are for the purpose of distinguishing the method of generating the electricity, but it is the same substance generated each time. As to its effects on the human system, if the exact truth be told, no person can be sure of what the real permanent effect is anyway, but there is no reason to suspect that merely the difference in the manner of generation would produce a difference in action. Any one of the ways could be used to generate electricity to kill a person, if proper arrangements were made. Likewise, electricity generated in any one of the three ways mentioned is perfectly harmless if passed through the body under proper conditions. Undoubtedly electricity generated chemically by means of a battery is most commonly used by the medical profession, but the reasons are the ease and convenience in handling and the fact that the potential or pressure of the electricity generated in the battery is more convenient for their purpose.

Just what electricity is no one knows, but the fact is not astonishing. No person knows what anything is. What is carbon? What is iron? What is oxygen? What is phosphorus? No one knows what any of these substances is yet, of course there are theories which explain in part. In the same way there is an electrical theory which is only of comparatively recent origin.

This theory states that everywhere throughout the universe, filling all spaces and all substances, there is an all-pervading material known as ether. It is this ether which transmits the light waves from the sun through the enormous distance between that heavenly body and our own atmosphere, which only extends a short distance above the earth. It is the ether which transmits heat from the incandescent filament within the vacuum bulb of an electric lamp to the glass itself and to the surrounding air. So electricity may be merely part of this ether in motion.

We do not need to know exactly the nature of electricity in order for it to be of value to us. With it we may light our houses, heat them, provide power for all purposes, perform chemical processes such as electroplating, and do multitudes of things of the greatest use and importance in the world's industry.

PART VI

USEFUL TABLES FOR ENGINEERING CALCULATIONS

I. The Equivalents of One Horsepower.
II. Absolute Efficiency of Various Engines.
III. Weights of Various Materials.
IV. Strength of Various Materials.
V. The Heating Value of Fuels.
VI. Water Heads and Corresponding Pressures.
VII. Water Powers for Various Heads.

TABLE I

THE EQUIVALENTS OF ONE HORSEPOWER

Power is the rate of doing work. A small engine, for example, can do as much work as a larger one but it will take a longer period of time. A four-horsepower gasoline engine will fill a silo satisfactorily in two days perhaps, but a twelve-horsepower engine will take only a little over half a day to do the same job. This illustrates the meaning of power. It has to do with time. We take as our unit of measurement the horsepower. That is the power which must be exerted to raise 33,000 pounds a distance of one foot in one minute. This is called exerting 33,000 foot-pounds of work in one minute. It is somewhat greater than most draft horses can exert continually. The following table gives not only the equivalent of one horsepower in foot-pounds for various units of time but also in foot-gallons of water and foot-cubic feet of water. A foot-cubic foot means that a cubic foot of water is raised one foot. A foot-gallon means that one gallon is raised one foot. The units given then mean that one horsepower will raise that number of cubic feet or of gallons a distance of one foot or half the number a distance of two feet, etc.

1 horsepower = 550 foot-pounds per second
 = 33,000 foot-pounds per minute.
 = 1,980,000 foot-pounds per hour.
 = 8⅘ foot-cubic feet of water per second.
 = 528 foot-cubic feet of water per minute.
 = 31,680 foot-cubic feet of water per hour.
 = 66 foot-gallons per second.
 = 3,960 foot-gallons per minute.
 = 237,600 foot-gallons per hour.

TABLE II

ABSOLUTE EFFICIENCY OF VARIOUS ENGINES

The mechanical efficiency of most engines is about 80 per cent. That is, the power given out by the engine is about 80 per cent. of that put into the engine. In a steam engine, for example, the power of the expanding steam exerted on the piston is the power put into the engine. About 80 per cent. of this is available for use at the belt pulley of the engine. There is another efficiency, however, which we may call the absolute efficiency. It is the proportion of the total power which could be exerted on the engine, that is given off at the belt pulley. In the steam engine, for example, only a portion of the expansive power of the steam is exerted on the piston. There is a tremendous loss in the exhaust steam. The same is true of the gasoline engine. There is a tremendous loss in the exhaust and in the heat radiated from the cylinder. In the windmill there is a loss due to the fact that some of the air goes through the wheel between the vanes and some of it exerts only a portion of its force on the vanes.

Absolute efficiency of any engine depends largely on the theory and design of that engine, while mechanical efficiency depends largely on the care used in manufacturing the engine and in keeping it oiled and running properly.

ABSOLUTE EFFICIENCIES

Steam engine	8 to 12 per cent.
Gasoline and kerosene engines	20 to 30 " "
Diesel oil engines	30 to 40 " "
Windmills	10 to 25 " "

Water-wheels
  Pelton . . . . 80 to 85 " "
  Overshot . . . . 65 to 75 " "
  Undershot . . . . 25 to 45 " "
  Breast . . . . 55 to 65 " "
  Turbine . . . . 55 to 85 " "

TABLE III

WEIGHTS OF VARIOUS MATERIALS

The specific gravity of any substance is the number obtained by dividing the weight of a cubic foot of that substance by the weight of a cubic foot of water. To find the specific gravity experimentally weigh the substance in air as usual. Call that weight W. Now weigh the substance while it is immersed in water. This can be readily done by tying a string to it and suspending it by the string from a spring balance, the substance being covered with water as by immersing it in a pail. Call this weight M. Then

$$\text{the specific gravity} = \frac{W}{W-M}$$

If the body is lighter than water and so floats, you must tie a sinker to it after having found the specific gravity of the sinker by the above method. Then let

    S = specific gravity of sinker
    K = weight of sinker in air
    N = weight of light body in air
    A = weight of the two bodies together in air
    R = weight of the two bodies together in water

and the specific gravity of the light body will be

$$= \frac{N}{A - R - \dfrac{K}{S}}$$

To find the percentage of different metals in an alloy or compound, find from the tables the specific gravity of the individual metals. Call these S and R. Get the weight (W) of the compound in air and its weight (M) in water. Then the amount of the metal whose specific gravity is S equals

$$\frac{W - R(W - M)}{1 - \dfrac{R}{S}}$$

For example, a knife is part silver and part steel. It weighs six ounces (0.375 lbs.) in air and 5½ ounces (0.343 lbs.) in water. The specific gravity of gold used is 15.7 and of steel is 7.8. The amount of gold in the knife is, therefore,

$$\frac{0.375 - 7.8(0.375 - 0.343)}{1 - \frac{7.8}{15.7}} = 0.26 \text{ lbs. of gold}$$

METALS

Name	Pounds per cubic foot	Name	Pounds per cubic foot
Aluminum	161	Lead	710
Carbon	216–222	Mercury	845.7
Copper	549–558	Nickel	540–550
Gold	1200	Platinum	1320–1350
Iron		Silver	650–661
Wrought	487–492	Tantalum	650–800
Cast		Tin	360–455
Gray	439–445	Tungsten	1160–1190
White	473–482	Zinc	439–449
Steel	474–487		

WOODS

Name	Pounds per cubic foot	Name	Pounds per cubic foot
Alder	26–42	Pine, Pitch	52–53
Apple	41–52	Hickory	37–58
Ash	40–53	Lignum vitæ	73–83
Bamboo	19–25	Linden	20–37
Beech	43–56	Locust	42–44
Birch	32–48	Maple	39–47
Butternut	24	Oak	37–56
Cedar	30–35	Pear	38–45
Cherry	43–56	Plum	41–49
Cork	14–16	Poplar	22–31
Ebony	69–83	Sycamore	24–37
Elm	34–37	Walnut	40–43
Pine, White	22–31	Willow	24–37
Pine, Yellow	23–37		

STONES

Name	Pounds per cubic foot	Name	Pounds per cubic foot
Brick	87–137	Limestone	125–190
Cement		Marble	157–177
Loose	72–105	Masonry	116–144
Packed	115	Mortar	109
Set	168–187	Mud	102
Coal		Sand	
Soft	75–94	Dry	87–103
Hard	87–112	Damp	119–128
Earth		Sandstone	124–200
Dry	100–120	Slate	162–205
Granite	125–187	Soapstone	162–175
Gravel	94–112	Trap	162–170

MISCELLANEOUS

Name	Pounds per cubic foot	Name	Pounds per cubic foot
Asbestos	125–175	Lime	144–200
Asphaltum	69–94	Paper	44–72
Bone	106–125	Paraffine	54–57
Butter	53–54	Peat	52
Charcoal	17.5–35	Rubber	57–62
Clay	122–162	Snow	7.8
Glass	150–175	Sugar	100
Ice	55–57	Tile	87–143
Leather	54–64		

LIQUIDS

Name	Pounds per cubic foot	Name	Pounds per cubic foot
Alcohol	49.4	Linseed oil	58.8
Alcohol, Wood	50.5	Lubricating oil	56.2–57.7
Benzine	56.1	Sulphuric acid	114.8
Gasoline	41–43	Water	62.4
Glycerine	78.6	Sea water	64
Milk	64.2–64.6		

TABLE IV

STRENGTH OF VARIOUS MATERIALS

There are three strengths which a piece of any material has. These are resistance to tension or stretching, resistance to compression or crushing, and resistance to shear or sliding of one layer of material past another layer in the same piece. When we tear a sheet of paper we really break it in shear by ripping one part of the paper sideways by the other part. The effect of shearing is exactly the same as though the piece of material were cut along the line of shear.

The tables given below are merely for guidance. No two pieces of any material will test exactly the same. In all work allowance is made for that fact and for the additional fact that we have no certain knowledge of just how the material will be strained, and so we must make the piece, whatever it may be, plenty large enough to resist all possible loads. To make this allowance, we use a "factor of safety." That is, we divide the ultimate strengths given below by 6 or 10 so as to give a safe working strength. These numbers by which we divide are the factors of safety.

Material	Tensil strength	Compressive strength	Shear
	In every case the number below means pounds per square inch		
Aluminum wire	30,000–40,000	......	......
Brass wire	50,000–150,000	......	......
Copper wire (hard drawn)	60,000–70,000	......	......
Platinum wire	50,000	......	......
Silver wire	42,000	......	......
Gold wire	20,000	......	......
Steel wire	460,000	......	......
Steel	80,000–330,000	56,000–70,000	48,000–60,000
Iron			
Cast	13,000–33,000	80,000	18,000
Wrought	50,000–54,000	46,000	40,000
Copper			
Cast	60,000–70,000	40,000	30,000
Tin	4,000–5,000	6,000	......
Zinc	7,000–13,000	20,000	......
Aluminum	15,000	12,000	12,000
Brass			
Cast	24,000	30,000	36,000
Lead	2,000	......	......
Rope			
Manila	9,000	......	......
Hemp	8,000	......	......
Leather	4,000	......	......
Granite	600	15,000	......
Limestone	1,000	7,000	......
Marble	700	8,000	......

Material	Tensile strength	Compressive strength	Shear
In every case the number given below means pounds per square inch			
Sandstone	150	5,000	......
Slate	10,000	10,000	......
Brick	200	10,000	......
Brickwork (lime mortar)	40	1,000	......
Brickwork (cement mortar)	300	2,000	......
Cement (Portland)	500	3,000	......
Concrete (Portland)	400	2,000	......
Oak (with grain)	10,000	7,000	800
Oak (across grain)	2,000	2,000	4,000
White pine (with grain)	7,000	5,500	400
White pine (across grain)	500	800	2,000
Georgia pine (with grain)	12,000	8,000	600
Georgia pine (across grain)	600	1,400	5,000
Cypress (with grain)	6,000	6,000	400
Cypress (across grain)	500	600	2,500

For compression and shearing use a factor of safety of 6. For tension use a factor of safety of 10.

TABLE V

THE HEATING VALUE OF FUELS

The heating value of any fuel is measured by engineers in units known as British Thermal Units, commonly abbreviated to the form B.T.U. A B.T.U. is the amount of heat required to raise the temperature of one pound of water through one degree Fahrenheit, the common temperature scale on the thermometer.

The following table gives the heating value of one pound of each fuel, and to get the heating value of any quantity such as a ton, multiplication must be made by the number of pounds in a ton (2,000 lbs.).

B. T. U. PER POUND OF VARIOUS FUELS

Fuel	Heating value in B. T. U.
Lignite (brown coal)	
Low grade	6,347
High grade	7,189
Bituminous (soft coal)	
Low grade	10,958
High grade	14,134
Semi-bituminous	
Low grade	14,121
High grade	14,699
Anthracite (hard coal)	
Low grade	12,577
High grade	13,351
Peat (Dried, matted, earthy matter)	From 8,761 to 10,307
*Gasoline (Sp. Gr. 0.71 to 0.73)	From 19,980 to 20,520
*Kerosene (Sp. Gr. 0.79 to 0.8)	From 19,800 to 20,160
*Alcohol (Sp. Gr. 0.82)	From 11,592 to 11,646

---

*Notice that these values are for a pound of the fuel oil, not for a gallon. The heating values per gallon may be calculated from the specific gravity of each and the weight of a gallon of water (8.34 lbs.).

TABLE VI

WATER HEADS AND CORRESPONDING PRESSURES

A column of water one foot high will exert a pressure of 0.433 pounds on every square inch of its base. In order to get a pressure of one pound per square inch, the column must therefore be 2.3 feet high. The following table gives the head in feet, that is the distance from the surface of the water to the point where the pressure is observed, and the pressure per square inch at the given distance below. It must be remembered that the size of the column does not affect in any way the pressure per square inch but, obviously, if there are more square inches to the column the total pressure will be greater.

Head in feet	Equivalent pressure in lbs. per sq. in.	Head in feet	Equivalent pressure in lbs. per sq. in.
1	0.433	10	4.33
2	0.87	15	6.45
3	1.30	20	8.66
4	1.73	25	10.78
5	2.17	30	12.99
6	2.60	35	15.16
7	3.03	40	17.32
8	3.46	45	19.49
9	3.90	50	21.65

Head in feet	Equivalent pressure in lbs. per sq. in.	Head in feet	Equivalent pressure in lbs. per sq. in.
55	23.82	100	43.30
60	25.98	150	64.95
65	28.15	200	86.60
70	30.31	250	108.25
75	32.48	300	129.90
80	34.64	350	151.55
85	36.81	400	173.20
90	38.97	450	194.85
95	41.14	500	216.50

TABLE VII

WATER POWERS FOR VARIOUS HEADS

A horsepower, as explained in Part III, Chapter XVII, is the equivalent of 33,000 foot-pounds per minute. Therefore the horsepower of falling water will be given by multiplying the number of pounds weight of water that falls in one minute by the distance that it falls, and dividing this product by 33,000. This horsepower is the theoretical amount and corresponds to the total energy contained in steam admitted to a steam engine. However, just as only a part of the energy in the steam can be utilized because the exhaust steam contains some of the heat energy, so the water which is expelled from the water-wheel contains some of its energy. The theoretical horsepower is therefore too large and it is usual to take 80 per cent. of that theoretical value for the practical amount which may be obtained. This percentage varies with the head and with the type of wheel as given in Table II.

The values given below are for the fall of one cubic foot of water per minute, and the weight of this is taken to be 62.4 pounds from Table III. If larger quantities are available, multiply the horsepower given by the number of cubic feet available. The number of cubic feet falling per minute is obtained by multiplying the cross-section of the stream (width in feet times average depth in feet) by the velocity of flow in feet per minute.

Head in feet	Practical horsepower	Head in feet	Practical horsepower
1	0.0015	75	0.1140
2	0.0030	100	0.1520
3	0.0046	150	0.2280
4	0.0061	200	0.3040
5	0.0076	250	0.3800
10	0.0152	300	0.4560
20	0.0304	350	0.5320
30	0.0456	400	0.6080
40	0.0608	450	0.6840
50	0.0760	500	0.7600

INDEX

A

Absolute efficiency, 207
Air, 162
Alcohol, 211, 215
Alum-soap, 31
Anthracite, 215
Asphalt, 31, 33
Asphaltum varnish, 44

B

Bacteria, 89, 93
Barbed wire, 53
Batteries, 154, 199
Battery rating, 151
Bituminous coal, 215
Blocks, Concrete, 33
Boiler, Steam, 99, 127
Brake horsepower, 126
Breast wheel, 133, 135
Brick, 33, 211, 214
British thermal unit, 214
Buildings, 3
Building material, 12, 14, 210, 211, 213, 214

C

Capillary action, 31, 160, 162
Catch basins, 171
Charcoal, 78

Circulation, 8, 13, 23
Coal, 98, 215
Coil
  Spark, 112
  Vibrating, 113
Cold storage, 21
Combustion, 115
Composition flooring, 36
Compressive strength, 212
Concrete, 26, 214
  Blocks, 33
  Composition, 40
  Laying, 30
  Lean, 27
  Non-absorbent, 31
  Portland, 29
  Proportions, 28
  Puddling, 30
  Theory of, 26
  Weight, 42
Conductors, Lightning, 48, 52
Cone of Depression, 61
Copper 52, 53
Costs of
  Batteries, 152
  Drainage, 168
  Fuel, 99
  Horse work, 110
  Irrigation, 183, 185
  Plowing, 98, 109
  Ram, 88

Costs of
  Ram-pump, 88
  Road building, 190
  Road maintenance, 192
  Tile, 165, 166
  Tractor, 105, 109
  Water, 185
  Waterpower, 141, 142, 144
  Water systems, 71
  Waterproofing, 32, 33
Cultivation, 163, 178
Cypress, 14, 214

D

Dam, 140
Deflectors, 24
Ditches, 174, 177, 181
Drainage, 159
  Ditch, 165
  Tile, 172
Drainage of buildings, 12, 17
Dressing, Top, for ice, 18
Driers, 44
Dynamic electricity, 200
Dynamometer, 124

E

Edison battery, 146, 148
Efficiency, 10, 83, 137, 146, 206
Electric battery, 145
Electricity, 47, 199
Electric theory, 201
Emery wheels, 40
Energy, 120

Engine, 98, 127
  Oil, 100, 116
  Steam, 100
Ether, 202
Explosion, 117, 118
Explosive mixture, 116

F

Factor of safety, 212
Felt, 34
Fertilizing, 164
Filter bed, 92
Filter, Sand, 75
Fireproofing, 36
Flooring, 36
Foot-pound, 120
Foundation, 15
Fuel, 98, 214
Furrows, 180

G

Galvanic electricity, 200
Gasoline, 97, 215
Gravel, 30
Gravity, Specific, 208
Gravity wheels, 134
Grindstones, 40
Ground connections, 54
Ground water, 159
Grout, 33

H

Hard water, 79
Head of water, 132, 139, 216
Heaters, Orchard, 194
Heating value of fuels, 214
Horse work, 103

INDEX

Horsepower, 119, 132, 205
  of boilers, 127
Hydraulic ram, 82
Hydrostatic water, 159

I

Ice
  Packing, 19
  Size of cakes, 14
  Top dressing for, 18
Icehouse, 9, 11
  Size of, 13
Ignition systems, 111
Impulse wheels, 134
Indicator, 121, 122
Insulation, 13, 16, 22
Insulators, 56
Irrigation, 178
  Skinner system of, 179, 182

J

Jump-spark ignition, 111, 113

K

Kerosene, 97, 215

L

Lamps, 151
Laterals, 175
Lead-plate batteries, 146
Lever, 125
Lighting systems, 149
Lightning rods, 47
Lignite, 215
Linseed oil, 43

M

Macadam road, 191
Magneto, 114
Maintenance, Road, 191
Make and break ignition, 111, 112
Material, Building, 12, 14, 210, 211, 213, 214
Mechanical efficiency, 206
Medical battery, 201
Metals
  Strength of, 213, 214
  Weights of, 210
Miner's inch, 131
Mixture, Explosive, 116
Moisture, 23

N

Non-conductors, 22

O

Operation, Costs of
  Engine, 99
  Horse, 110
  Tractor, 108
Orchard heaters, 194
Outlets, Drainage, 174, 175
Overshot wheel, 133, 134
Oxychloride binders, 37

P

Packing
  Ice, 19
  Stores, 25
Painting, 3, 43
Paint spraying, 45
Paper, Roofing, 15, 16, 22

Paraffin, 32
Peat, 215
Pelton wheel, 133, 135, 136, 143
Planté plates, 147
Plow gang, 107
Pneumatic equipment, 70
Potential, 199
Power, 120, 129, 205, 217
Pressure, Water, 216
Prony brake, 124
Protection of orchards, 195
Public roads, 190

Q

Quicksand, 176

R

Radiation, 5
Rain water, 62, 66
Ram, 82
Ram-pump, 82
Rating of
  Battery, 151
  Boiler, 127
  Engine, 127
Red lead, 44
Road
  Building, 190
  Cost of, 190, 192
  Maintenance, 191, 192
Rods, Lightning, 47
Roofing paper, 15, 16, 22
Roofs, 17
Round buildings, 4
Running water systems, 68

S

Sand, 26, 29
Sand-clay roads, 191
Sand filter, 75
Seepage, 160
Septic tank, 90
Sewage, 89
Shape of buildings, 4
Shear, 212
Shell road, 191
Sills, 17
Silo, 6, 8
Size of icehouse, 13
Skinner irrigation system, 179, 182
Spark, 117, 198
Spark coil, 112
Specific gravity, 208
Spraying paint, 45
Springs, 65
Static electricity, 199
Steam boiler, 99, 127, 146
Stones, 26, 29, 211, 214
  Artificial, 36
  Beton-Coignet, 41
  Ransome, 41
  Sorel, 36
Storage battery, 145
Storage, Cold, 21
  Packing, 25
Strength of materials, 212
Sylvester treatment, 31, 35

T

Tamping, 19
Tensile strength, 212
Tile, 17, 165, 166
  Laying, 173

Tiling systems, 169
Time-saving, 4
Thermal unit, 214
Top dressing for ice, 18
Tractor, 102
Trap, 17
Truck, 111
Turbines, 137, 138, 142
Turpentine, 43

U

Ultimate strength, 212
Undershot wheel, 133, 135

V

Varnish, 44, 46
Ventilation, 9, 12, 25
Ventilators, 18
Voltage, 148, 152

W

Water
  Flow, 184
  Head, 216
  Measurement, 206
  Power, 129, 217
  Supply, 61
  Table, 61
  Wheels, 133
Waterproofing, 26, 36
Weights, 208
Wells
  Artesian, 65
  Deep, 64
  Dug, 63
Whitewash, 45
Woods
  Strength of, 214
  Weights of, 210
Work, 120, 124

V. N. D.

THE COUNTRY LIFE PRESS
GARDEN CITY, N. Y.

CPSIA information can be obtained
at www.ICGtesting.com
Printed in the USA
BVHW040920101218
535233BV00020B/745/P